Good Science

Good Science

The Pursuit of Truth and the Evolution of Reality

Timothy McGettigan

LEXINGTON BOOKS
Lanham • Boulder • New York • Toronto • Plymouth, UK

Published by Lexington Books
A wholly owned subsidiary of The Rowman & Littlefield Publishing Group, Inc.
4501 Forbes Boulevard, Suite 200, Lanham, Maryland 20706
http://www.lexingtonbooks.com

Estover Road, Plymouth PL6 7PY, United Kingdom

British Library Cataloguing in Publication Information Available

Library of Congress Cataloging-in-Publication Data

McGettigan, Timothy.
Good science : the pursuit of truth and the evolution of reality / Timothy McGettigan.
p. cm.
Includes bibliographical references and index.
ISBN 978-0-7391-3677-5 (cloth : alk. paper)
1. Truth. 2. Science—Philosophy. I. Title.
BC171.M37 2011
001—dc23 2011026236

The paper used in this publication meets the minimum requirements of American National Standard for Information Sciences—Permanence of Paper for Printed Library Materials, ANSI/NISO Z39.48-1992.

Printed in the United States of America

Dedicated to Karl Popper and the many problem-solvers that he has inspired.

Contents

Foreword

It's easy to beat up on postmodernists these days. Ever since the Sokal Hoax, the postmodernists' Waterloo, the science wars have been a rout. Once it became clear that postmodernism was incapable of distinguishing between valid scientific perspectives and gibberish, postmodernists have bolted from the battlefield.

This was a remarkable turnabout. For much of the 1990s, postmodernists insisted that modernist science was as good as dead. Modernists, you must understand, included scientists and anyone else who was naïve enough to believe that science systematically produced better, more enlightening knowledge.

Not so, claimed postmodernists. Under the guise of truth-seeking, postmodernists contended, science had woven a worldwide web of deceit. Certainly, some scientists might have been earnest believers in admirable principles; however, the net effect of scientific progress did little more than aggrandize the West at the expense of the downtrodden. And the prime culprit for those dark deeds was none other than the concept of truth.

Whereas scientists tended to view truth as a benign standard against which to gauge scientific progress, postmodernists argued that truth was an evil instrument of cultural discipline. Ideas which complied with Western truth standards merited approval, whereas ideas that challenged the modernist truth regime were subjugated and marginalized.

As a remedy, postmodernists advocated the end of truth. In this way, Western bias could be dethroned and all of the ideas that had been marginalized by modernism would finally get a fair hearing. Of course, as Sokal illustrated, if we abandon truth, we also abandon rationality. In a world of postmodern relativism, anything goes. Without truth standards, there is no way to distinguish between good and bad ideas: all ideas are equally valid. Which is a really bad thing—unless you're convinced that Hitler and Stalin were visionaries.

I must admit, I was perfectly happy to see postmoderism implode. Postmodernism was a gutless theoretical movement that arrogated unto itself the right to criticize everyone else's ideas while failing to produce any worthwhile ideas of its own. The best part was that, throughout its meteoric rise, postmodernism constantly propounded the imminent demise of modern science. To that, all I can say is "Ask not for whom the bell tolls . . . " Clearly, it would have been more accurate for postmodernists to predict their own demise. However, accuracy was never a priority among postmodernists.

So, where does that leave us?

Irksome as it may have been for postmodernists, science has forged ahead—before, during and after the postmodern interlude—with nary a hiccup. The biggest threat to science during the past couple of decades was the Bush Administration. Compared to Dubya, postmodernism was like a gnat on a water buffalo's backside.

If we can thank postmodernism for anything, it is for refocusing attention on the knotty issue of truth in science. Rightfully, Karl Popper should get most of the credit for problematizing the concept of truth. Of course, Popper took a much different view of the role that truth should play in humanity's never-ending problem-solving endeavors. Still, Popper made it clear that truth was not nearly as straightforward a phenomenon as most scientists, particularly positivists, liked to think.

Thus, our story.

The question that I will address in this book is: Precisely, what is the relationship between truth and science—and *why does it matter*? In brief, I believe that the pursuit of truth is the most important endeavor that humans have ever undertaken. In seeking the truth, humans have regularly and profoundly redefined reality for the better. If you're happy to be living in a *post*-postmodern world, thank a scientist. Yet, stimulating as life in the information society may be, the world remains full of problems. Further, if anything, those problems tend to grow more challenging by the year. The 2008 global financial meltdown came within a whisker of ending civilization as we know it.

In the pages that follow, I will argue that truth-seeking is not just the surest way to build a better, brighter future for everyone, but it's the only way to create a future that's worth living for.

So, why does truth matter? I sincerely hope an answer to that all-important question will soon become evident as you turn the pages of this book.

Acknowledgments

I would like to thank an anonymous reviewer for providing insightful comments that helped improve this book considerably. Also, I would also like to thank Susan, Claire and Ruby for being the best partners that I could ever hope for in life's long, pleasant journey. The best is yet to come.

Introduction
What is Good Science?

"I am become Death, the Destroyer of Worlds" —Robert Oppenheimer

Science is a vehicle for change. Arguably, no human endeavor has ever altered the course of history more dramatically than science. Further, the effects of scientific inquiry on the earth and its inhabitants, though already substantial, are certain to increase in the future. Homo sapiens is fast becoming Homo scientia.

Put simply, science can be understood as the pursuit of knowledge. Though scientists come in many varieties and they study an extraordinary diversity of subject matter, there are certain things that all scientists have in common. Scientists are knowledge-seekers, or, more accurately, good scientists are truth-seekers. Trifling as it may seem, that is actually a crucial distinction. Knowledge comes in many forms. Virtually any idea that pops into someone's head (i.e., from pink unicorns to space elevators) can be considered a form of knowledge. However, out of all the knowledge that humans generate, only a tiny fraction can be considered truthful. So, that begs the question, "What is truth?"

Truth is a challenging subject. Many people, whether they describe themselves as scientists or not, might insist that truth is nothing more than a description of facts. For example, it is true that the sun rises in the morning and sets in the evening. At first glance, such a definition seems perfectly reasonable: truth *should* correspond with facts. However, the danger of such a definition is that facts are not always what they seem. Take, for example, the fact that the sun rises and sets on a daily basis. Although that statement offers a plausible description of certain facts, nevertheless, it is not true. The sun does no such thing. Based upon what astronomers have learned over the past several centuries, we know that the sun does not orbit the earth. Instead, the earth's rotation tends to instill the false impression that the universe revolves around earthlings. Our real relationship with the cosmos is very different. Additionally, for those who dwell near the earth's poles, rather than rising and setting on a daily basis, the sun often appears and disappears for months at a time. Consequently, facts often look very different depending upon one's perspective.

That said, it is important to emphasize that there is an essential relationship between truth and facts. In other words, one can't say anything truthful without reference to verifiable facts. Thus, I might claim that I have spotted the Sasquatch in my backyard; however, unless I can produce hard evidence of such a mythical visit, no one should believe a word I say. Good scientists certainly wouldn't.

Typically, good science can be understood as knowledge-seeking activities that assert a very clear linkage between truth and facts. Thus, for the most part, good scientists tend to view ideas that are not supported by facts (e.g., Sasquatches popping in for tea) as fantasies. Indeed, imaginative humans have a penchant for dreaming up all sorts of notions that, scientifically speaking, are rubbish. Consequently, good scientists usually draw a sharp distinction between facts and fantasies. Good science is devoted to the former and dismissive of the latter.

In important respects, this perspective is entirely justifiable. Facts matter. However, I will argue in this book that scientific progress is often contingent upon seeking truths that lie beyond established facts. In other words, fantasies can often inspire scientific progress that facts might otherwise impede. In chapter one, I will examine a critical juncture in the history of science: the trials and triumphs of Galileo Galilei. Good scientists often pay a hefty price for advancing new truths. Nowhere is this better illustrated than in Galileo's persecution at the hands of the Inquisition.

As a result of his astronomical observations, Galileo discovered important new facts in the heavens. Unfortunately, for Galileo, his discoveries could not be reconciled within the leading cosmology of his day. Rather than embracing Galileo's novel ideas, powerful forces repudiated Galileo for challenging sacred truths and, implicitly, the authority of the dominant theological structure. In the end, Galileo was forced to recant, and good scientists the world over learned an important lesson: knowledge and power are inextricably linked. May truth-seekers beware.

In chapter two, I will focus on the relationship between power and knowledge. Against the backdrop of Charles Darwin's efforts to develop his theory of evolution, I review the legendary debate between Karl Popper and Thomas Kuhn. While Popper was convinced that truth was indispensable to the endeavors of good scientists, Kuhn dismissed truth as a chimera. For Kuhn, truth was an inconsistent standard that drifted to and fro with every paradigmatic shift.

Drawing upon the strengths of each argument, I will advance an alternate definition of truth. As Darwin's evolutionary intellectual development illustrates, although facts may appear dramatically different from the perspective of one paradigm or another, truth is a consistent standard that enables good scientists, such as Charles Darwin, to develop entirely new explanations for previously inexplicable facts.

In chapter three, I examine the complexities of defining truth within the context of multi-dimensional social realities. Robert Dahl argued that good scientists should confine their inquiries to empirically observable facts. On the other hand, C. Wright Mills contended that most social phenomena are invisible to the eye and, thus, can only be imagined. I argue that good science needs to be redefined in order to take the invisible, but very real, influences of invisible social phenomena into account. Doing so not only broadens the definition of social reality, but also redefines the nature of truth-seeking.

In chapter four, I examine the influences of scientific fact production on the evolution of new realities. In the late 1950s, the United States and Soviet Union embarked on the space race. The early stages of the space race were marked by a seemingly endless series of Soviet triumphs. In an effort to boost the United States out of its technological doldrums, President John F. Kennedy challenged his countrymen to resolve a seemingly impossible problematic: landing astronauts on the moon by 1970. A problematic can be understood as a daunting intellectual challenge that can only be resolved by redefining reality.

In other words, when JFK announced his moon-landing problematic, the United States lagged behind the Soviets in every critical area of space science achievement. Thus, for the United States to beat the Soviets to the moon, the United States would need to become the world's most technologically-advanced nation. Far-fetched as it may have seemed, the United States had to invent the facts that would transform Kennedy's fantasy into a reality. In doing so, the United States reinvented itself as the most powerful, and technologically-advanced nation on the planet. Thus, new truths are sometimes more a product of fantasy than reality.

In chapter five, I will examine a contemporary, and as yet unresolved, problematic: artificial intelligence (AI). In 1950, Alan Turing proposed that computers would one day rival the intelligence of their human creators. In many ways, Turing's challenge has inspired the production of entirely new facts (i.e., new types of smart machines), and, in so doing, Turing's problematic has helped to transform the substance of social reality. While I disagree with researchers who argue that AI already exists, nevertheless, I believe that Turing's problematic will continue to inspire significant technological achievements for some time to come.

In chapter six, I will review the celebrated debate between Albert Einstein and Niels Bohr. Although Einstein made crucial advancements in the field of sub-atomic physics, Niels Bohr ushered quantum science into an entirely new realm. In brief, Bohr's Copenhagen interpretation asserted that quantum reality was an uncertain domain wherein particles existed only in connection with human observers. Though Einstein found it difficult to stomach much of Bohr's reasoning, no idea distressed him more than the proposal that quantum mechanics was a complete theory. Although experimental tests have consistently supported the mind-bending tenets of quantum mechanics, Einstein remained an opponent of the theory to the bitter end. While Einstein's opposition may have damaged his stature among fellow physicists, nonetheless, his principled objections to quantum mechanics offer evidence of Einstein's unremitting commitments to good science.

In closing, I offer a few thoughts about the future. Although the world is plagued by seemingly insoluble problems, crises have always been endemic to human civilization. I argue that the clearest path to the brightest future will be to pursue the most challenging problematics that knowledge-seekers can invent and, thereby, commit the human race to a never-ending process of redefining reality.

Chapter 1
Good Science in a Dark Age

During the medieval era, Europeans clung to a number of quaint beliefs about the structure of the universe. For starters, Dark Age Europeans believed that the universe was relatively small, young and, most important of all, in shallow orbit around the earth. From the vantage of the twenty-first century, such notions seem inexcusably rustic. Nevertheless, medieval cosmology was actually based upon compelling empirical evidence. Without the assistance of telescopes, the entire celestial vault appears to arc dutifully overhead. In addition to describing the mechanical relationship between the earth and the heavens, naked eye observations also gave rise to a full-blown earth-centered cosmology: geocentrism.

Cosmologies are complex belief systems that provide an intelligible framework within which to make sense of the universe. As a cosmology, geocentrism asserts that not only does the entire universe revolve around the earth, but the earth is literally the most important object in creation. During the Middle Ages, geocentrism operated hand in glove with Europe's predominant Christian ideology. In other words, Biblical theology concurred about the earth's centrality, while it also imposed a ranking system on every object in the cosmos: The Great Chain of Being (Lovejoy, 1936).

According to The Great Chain of Being, the universe revolves, both literally and figuratively, around God. As the logic goes, God would naturally want to keep his most precious creations nearby and, thus, he conferred upon humans the unique privilege of residing on the earth; the most central and, thus, most important location in the universe.

Just as The Great Chain of Being ranked heavenly objects into a fixed hierarchy, a similar logic also applied to earthly phenomena. That is, the people who occupied the highest ranks of various social strata (e.g., kings, queens, popes, etc.) were considered to hover somewhat closer to God's sacred realm. As a result, The Great Chain of Being supplied a convenient justification for the sanctity of entrenched political hierarchies. The divine right of kings and the ideological domination of the Church were sustained by the undeniable symmetry of the universe. Elites had justifiably greater access to power because they enjoyed a more intimate relationship with God. To challenge geocentrism would not only defy commonsensical explanations of earth's relationship to the cosmos, but it would also imply a heretical denunciation of the ruling ideology. Consequently, stargazers who presumed to cast doubt on the earth-centered structure of the universe would only do so at grave peril.

Enter Galileo

Galileo Galilei (1564-1642) is considered to be one of the world's first and greatest scientists (Sobel, 2000). What made Galileo unique was his ability to observe carefully, develop original theories, and test his ideas systematically. Thus, Galileo embraced the doctrine of good science long before science coalesced into an institutionalized set of principles and practices.

Following the invention of the first telescope in 1608 (King, 1955), Galileo took up this new instrument with uncommon zeal. Imperfect as early telescopes may have been, Galileo was nonetheless enthralled by the wonders that his new star-gazing tool exposed. One glance through the telescope revealed that there was much more to the heavens than anyone had previously imagined: more stars, more planets, more beauty and more anomalies. Indeed, Galileo is credited with making a plethora of astronomical discoveries that served to both delight and alarm his contemporaries. Ultimately, Galileo's most controversial discovery resulted from his scrutiny of the planet Jupiter.

Following repeated observations, Galileo was astounded to discover that there were objects that moved in a steady circuit around Jupiter. Those objects were faint enough that, even on the clearest of nights, it would have been impossible to discern their existence without the aid of a telescope. However, with the help of his new stargazing instrument, the objects became clear enough to convince Galileo that he had witnessed a wondrous new truth: Jupiter had moons.

To put it mildly, Galileo's discovery represented a major scientific breakthrough. No earth-dweller had ever observed satellites in orbit around another world. Although one might suppose that Galileo's revelation would have inspired widespread celebration, in fact, Galileo's landmark discovery invoked the ire of a truly lethal adversary. Whereas Galileo may have hoped to blaze a new path through the heavens, he would soon witness his work condemned as an unwelcome intrusion upon inviolable terrain.

Sowing the Seeds of Doubt

In the four hundred years since Galileo published his discoveries, conceptions of the universe have transformed remarkably. Whereas medieval stargazers once believed that the entire universe revolved in concentric circles around the earth, astronomers now theorize that the universe is literally billions of light years in diameter (Finkbeiner, 2010). Arguably, the vast changes that have redefined scientific cosmology can all be linked to the fallout from Galileo's discovery of Jupiter's moons.

Once again, Galileo lived at a time when the Catholic Church was Europe's foremost political power. The medieval Church comprised a vast international organization at a time when most other European political entities were no larger than city-states or feudal fiefdoms. As is the case with most theological systems, faith was the glue that bound the entire Catholic superstructure together. Expressions of faith not only defined the parameters of membership in the Catholic Church, but faith also delimited the boundaries of acceptable intellectual inquiry. For Catholics, The Truth was defined by the established parameters of Catholic dogma (e.g., God created the universe in six days). In other words, so long as the faithful restricted their intellectual inquiries to the limits prescribed by the Church, then such choreographed mental exercises would conveniently reproduce faith in Church doctrine.

For those who govern through the strictures of faith, doubt is equivalent to disobedience. Not only does doubt connote a certain degree of disrespect for sacred beliefs (i.e., "good" Catholics all had an obligation to embrace the sacred word of God), but doubt also weakens organizational structures that are founded upon blind obedience to faith. In other words, doubt implies that faith-based belief systems are somehow inadequate. As such, doubt also portends a quest for knowledge that lies outside the authorized realms of inquiry. By thinking outside the box, it becomes possible for doubters to generate novel observations that are not only distinct from, but that are often directly contradictory to established beliefs. As a result, free-thinkers tend to evoke antipathy among those who maintain a vested interest in the status quo.

For example, cognizant of the hostility that his radical new perspective was likely to inspire, Copernicus (Repcheck, 2007) literally delayed until he arrived on his deathbed to unveil the culmination of his lifework: a heliocentric theory of the universe. Quietly and systematically, Copernicus had nurtured his doubts about geocentrism until he managed to invent a more convincing theory of heavenly motion: Copernicus moved the sun to the center of the universe. Modest as this astronomical shift may have been, Copernicus was convinced that his sun-centered theory would invoke the ire of the medieval theological establishment (Owen, 2004). Though Copernicus' deathbed revelation might seem to over-dramatize the danger that he faced, in fact, Galileo would soon discover that his colleague's fears were well-founded.

Galileo was acquainted with Copernicus' heliocentric theory (Hofstadter, 2009), and he was also sensible of the reasoning behind its near-posthumous publication. Though more subtle, Galileo's discovery of Jupiter's moons posed yet another serious threat to geocentric theory. Once again, geocentric theory asserts that the entire universe revolves around earth. However, Galileo's observations suggested that there were in fact multiple centers in the universe (Finocchiaro, 1989). In such a universe, the earth would descend from lofty preeminence to a stature of banal equivalency with every other object in creation. As a result, the earth's rulers could no longer base their mandates on the logic of The Great Chain of Being. That is, if the earth was nothing special, then the same would also be true of its denizens. A demotion for the earth implied a similar demotion for its formerly exalted leaders.

Friends in High Places

Aware that he was skating on thin ice, Galileo sought advice from a number of key power-brokers before going public. Though it was clear that not every member of the Catholic hierarchy was enamored of Galileo's discoveries, in the end, the plaudits that he received, particularly those from Pope Urban VIII, convinced Galileo that it would be safe to publish (Sobel, 2000). Yet, in spite of Galileo's precautions, shortly after publishing a monograph that commented favorably upon heliocentrism, the Inquisition sent Galileo a summons (Blackwell, 2006).

In 1633, the Inquisition subjected Galileo to extended interrogation. During the course of this inquiry, Galileo affirmed that he had indeed advocated the plausibility of heliocentrism. Given that Galileo was a famous scholar who was also blessed with impressive political connections, the Inquisition was inclined to be lenient. Instead of handing down an outright conviction of heresy—which would have incurred a death sentence—the Inquisition offered Galileo an olive branch: Galileo could recant (Doak, 2005). Thus, Galileo was faced with two very unpalatable alternatives. Either Galileo could take a principled stand and affirm his heretical support for heliocentrism, or he could eat crow and renounce his former position. Further complicating matters, Galileo had more than his own life to consider. If he were to be convicted of heresy, the stigma of his sentence would also extend to his family (Sobel, 2000). Given the harrowing implications of a heresy conviction, Galileo decided that he had no choice but to recant. The truth be damned.

Thus, the Inquisition had won. Not only did Galileo publicly reverse his position on heliocentrism, but the Inquisition kept him under house arrest for the rest of his days. Anyone who could pose such a dire threat to geocentrism was best kept under close surveillance. Indeed, the watchful eye of the Inquisition and the ever-present threat of a renewed heresy charge produced the intended effect. As Galileo whiled away his remaining years, his scientific inquiries dwindled to nothing. Galileo was defeated and his ideas had been crushed by an overbearing ideology.

Or had they . . . ?

The Truth Be Damned

Though Galileo had been forced to recant, the Inquisition could not entirely extinguish his ideas (Steele, 2008). Galileo had sparked a tenacious fire. He had stated publicly that the universe did not fit within the artificial constraints of geocentric cosmology. Thus, Galileo illustrated that there was a larger universe to discover, but only for those who had the courage to seek the truth. Following Galileo's lead, stargazers defiantly tilted their telescopes skyward and delved deeper into the mysteries that had been repudiated by the high priests of geocentrism. It was an epic struggle. The very definition of the universe was at stake (D'Onofrio and Burigana, 2009).

Although the concept may have been unknown in medieval Europe, the struggle that ensued between Galileo and the Catholic Church can be understood as an example of paradigm crisis. Thomas Kuhn (1962) introduced the idea of paradigm crisis to explain the evolution of scientific thought. Instead of undergoing a smooth, linear progression, Kuhn argued that scientific thinking tends to be more disjointed. The antipathy generated by Galileo's heliocentric model of the universe illustrates that scientific transitions are usually more like collisions than collaborations.

Thomas Kuhn (1962) argued that scientific revolutions take place whenever dominant paradigms are dislodged by emergent paradigms. Paradigms can be understood as broad, assumption-laden world views that provide a theoretical framework into which knowledge-seekers can integrate facts and observations. Science undergoes paradigm shifts when established paradigms fail to account for an increasing number of empirical anomalies. Anomalies can be understood as enigmas for which existing knowledge systems lack convincing explanations. Kuhn's perspective challenged the previously accepted view that the accumulation of scientific knowledge was a rational stepwise process, i.e., each landmark discovery being anticipated with logical precision and, once established, elevated consensually atop a vertical tower of knowledge. Instead, Kuhn contended that paradigm shifts are much messier undertakings that are marked by infighting, political subterfuge, and a host of other unscientific antics. In other words, though scientists are generally loath to admit it, the accumulation of scientific knowledge is a social enterprise and is, thus, replete with human shortcomings.

Paradigms Unlimited

Varied as paradigms may be, they all share the quality of being flawed. Paradigms, being the product of imperfect human intellects, are but a reflection of those intellects: ingenious but fallible. Due to their deficiencies, Kuhn argues that paradigms tend to progress through fairly typical life cycles. In their early stages, paradigms enjoy broad support and relative stability. This is the case because paradigms are generally developed to explain previously unintelligible phenomena. Thus, early in their life spans, paradigms enjoy the advantage of being able to explain a wide variety of phenomena more effectively than pre-existing paradigms.

Nevertheless, in every case, given the flaws that inhere within any human-made knowledge framework, anomalies are bound to emerge. Thus, Galileo's observations of Jupiter's moons constitute an inexplicable anomaly for the geocentric perspective. After a time, anomalies begin to accumulate in sufficient numbers to threaten the viability of dominant paradigms. When this occurs, Kuhn argues, paradigms enter a crisis phase. Paradigm crisis is marked by the emergence of a competing paradigm that offers more logical explanations for the anomalies that threaten to undermine a dominant paradigm. Thus, not only did Galileo observe Jupiter's moons, but he also had the temerity to propose heliocentrism as a superior alternative to geocentrism.

During paradigm crisis, rather than systematically building bridges from an outmoded to a more advanced paradigm, intellectuals tend to become polarized. Intellectuals whose authority, status and privilege are linked to the ongoing dominance of a particular paradigm declare war against intellectuals who advocate an emergent paradigm—and who stand to gain new levels of authority, status and privilege if their ascendant paradigm manages to dislodge its better-established opponent. Due to such allegiances, the intellectual battles that ensue during paradigm crises are often more emotional than logical. Once again, rather than methodically weighing the evidence that Galileo had introduced in support of heliocentrism, the Inquisition threatened Galileo's life for slighting the sacred tenets of geocentrism. The Inquisition was not willing to consider logical alternatives to geocentrism, instead its sole purpose was to curtail perceived threats to its preferred ideology.

Interestingly, Kuhn argues that passions tend to run so high, and intellectual opponents become so fiercely polarized during paradigm crises that such intellectual contests become irresolvable. In the end, emergent paradigms gain prominence not due to their more compelling logic, but only due to the demise of their former adversaries. That is, new paradigms replace older paradigms only when a critical mass of young turks successfully dislodge the supporters of formerly dominant paradigms. As a result, instead of a smooth, linear progression from flawed to increasingly competent intellectual paradigms, Kuhn argues that the history of ideas is better understood as a disjointed journey from one set of mutually exclusive ideas to another.

Thus, in spite of the fact that the Inquisition successfully silenced Galileo, heliocentrism eventually supplanted geocentrism. However, this transition did not occur because the respective antagonists were able to resolve their differences. Indeed, it is worth noting that it was not until the year 1983 that the Catholic Church conceded that Galileo may have been correct (Brown, 1990, p. 385). Of course, in the realm of scientific endeavor, as of 1983, the veracity of Galileo's observations was, to put it mildly, a foregone conclusion.

It has been approximately four hundred years since Galileo aimed his new-fangled telescope skyward. In doing so, Galileo pointed the way to an entirely new universe. In part, Galileo changed the universe by discovering unknown objects in the firmament. More importantly, Galileo changed the universe by crafting a scientific alternative to the dominant cosmology of his day. In other words, medieval European cosmology was constrained by the ideological priorities of a self-serving ruling class. As such, discoveries that supported the dominant ideology were given a hearing, whereas ideas which posed a threat to that discourse were marginalized. In spite of such biases, Galileo proposed a scientific alternative to geocentrism. Galileo's subsequent travails signaled a paradigmatic changing of the guard. Whereas the Catholic Church dominated intellectual inquiry throughout the medieval era, beginning with Galileo, science has become the dominant instrument of truth.

From Darkness to Light

Indeed, in collaboration with numerous other thinkers, Galileo helped to initiate the Age of Enlightenment. The Enlightenment was an era during which Europe metaphorically moved from intellectual darkness into the light (Outram, 2005)—and the brightest beacon in this transition was science. From its inception, science has shed new light on every phenomena that has fallen under its gaze. From the orbits of heavenly bodies to the physics of sub-atomic particles, science has not only transformed the way that knowledge-seekers perceive reality, but it has also revolutionized the structure of lived reality. The fast-developing technologies that distinguish the modern world from its pre-industrial past represent practical applications of science. However, as was the case with the Wright Brothers' flying machine (Wright, 1998), the introduction of new technologies must be understood, very literally, as a transformation of reality. Whereas aviation was once ridiculed as a fantasy, in the post-Wright Brothers world, aviation has radically redefined the texture of lived reality (Shulman, 2003). Until recently, personal computing merited similar skepticism. As late as the 1960s, personal computers were the stuff of science fiction (i.e., "Beam me up, Scotty!"). Yet, in the space of only a few short decades, computers have shrunk from multi-ton mainframes to pocket-sized cyberspace plug-ins (Swedin and Ferro, 2007).

Given the global scope of the scientific revolution, one could argue that science is not only distinct from, but it is also superior to all other knowledge systems. From finding longitude (Sobel, 1995) to interplanetary exploration (Thimmesh, 2006), no other knowledge-production system can match the exploits of science. Yet, in spite of its successes, some have argued that science is simply the latest in a long history of self-serving ideological systems. Seizing upon Kuhn's ideas, postmodernists have asserted that science is nothing more than an instrument of global oppression.

A Recipe for Good Science

The central objective of modern, enlightened science has been to seek the truth (Lemert, 2004; McGettigan, 1999, 2000). Indeed, not only has science created the modern world, but it has also promised to create a better society. The basis for such a bold assertion was that science rested on a qualitatively stronger epistemological foundation than any other belief system. Whereas other ideological systems were regulated by dogmatism, science was beholden to no earthly influence save the most estimable of intellectual goals: the pursuit of truth.

Philosophically speaking, the possibilities of applying science to the construction of an evolving society were, to say the least, exciting. In an enlightened age, the scourge of social pathology would doubtlessly wither under the antiseptic light of scientific scrutiny. Further, sweeping democratic reforms promised to unite with science in a mutual struggle to uplift humanity. For it to flourish, good science required democratic freedoms that ensured unfettered opportunities for inquiry. In return, science promised to invigorate the public's long neglected intellect. Fueled by their reciprocal aspirations, science and democracy have indeed combined to profoundly influence the people, politics, and economics of the modern world.

Yet, despite the spectacular achievements of modernity's juggernaut (Giddens, 2010), the fabled good society remains unrealized. Certainly, there are many who enjoy luxurious living standards. Nevertheless, as modernity has progressed, the number of people living in abject poverty has also increased (McLaren, 2000). While some have insisted that culpability lies with those who are least fortunate (Murray, 1984, 1998; Rostow, 1960), others contend that fault lies elsewhere.

Postmodernists have hammered hard at the illusion propagated by modernists that science would ultimately generate truth and, thereby, a better world for one and all (Clough, 1994; Denzin and Lincoln, 1994; Kincheloe and McLaren, 1994; Lather, 1991; Lemert, 1991, 2004; Lyotard, 1984; Richardson, 1991; Seidman, 1991; Tierney, 1997). Indeed, if modernity was supposed to ameliorate social problems, then why, after approximately four hundred years of enlightening scientific development, has global misery actually *increased*?

Postmodernists answer: Because the Enlightenment's promises were lies.

Far from achieving the Enlightenment's philosophical goals, postmodernists maintain that science has sustained a long-standing project to exploit people everywhere (Denzin, 1997; Kincheloe and McLaren, 1994; Lather, 1995; Lemert, 1991; Seidman, 1991). In other words, the Enlightenment is a western European invention that is permeated by a variety of biases (e.g., patriarchy, racism, Christianity, rabid industrialism, etc.). Thus, despite pretensions of objectivity, Western cultural biases have thoroughly tainted every aspect of modern progress (Smith, 1990): the language, customs, and ideals of the West are emulated everywhere.

Postmodernists have pointed out that all knowledge is constructed within bounded socio-cultural systems and, whether characterized as truth or not, no form of knowledge could ever be more universal than the social system in which it was created. While this is a markedly different contention than than Kuhn's, nevertheless, these points resonate with his characterization of paradigmatic deficiencies: paradigms aspire to compass the "the known universe" but invariably fail because they impose limited epistemological frames on a boundless universe. Postmodernists advance the additional criticism that, not only does knowledge tote the baggage of the social context in which it was generated, but it also imposes ideological coercion upon all those who are exposed to it (Lemert, 1991). Proceeding from those assumptions, postmodernists conclude that the modern, scientific regime of truth is nothing more than a covert justification for Western imperialistic abuse (Seidman, 1991; Lemert, 2004).

As an antidote, postmodernists decided to jettison the notion of universal truth in favor of embracing individual-level truths. This diminution of truth standards presumably remedies the gravest injustice perpetrated by modern science: exclusionary truth standards have invalidated the voices of the powerless. Postmodernists endeavored to rectify this problem by asserting that all knowledge is equal.

While abandoning universal truth purports to serve a noble purpose, nevertheless, doing so also creates a number of drastic epistemological dilemmas (Sokal, 1996, 2010). The first of these is relativism: by abandoning truth one foregoes all non-relative means of evaluating knowledge. In other words, in a universe wherein there are no supra-individual truth standards, any knowledge claim by any individual (including luminaries such as Adolf Hitler and Joseph Stalin) must be treated as equally valid. Postmodernism establishes no basis upon which to develop broader evaluations of truth (McGettigan, 2000, 2008). Ergo, in the absence of universal truth standards, one cannot avoid endorsing the machinations of tyrants.

Largely due to the irresolvable epistemological dilemmas that it evokes, postmodernism has fallen out of favor. Nevertheless, the Kuhnian critique of science remains a subject of active debate. Given the often contradictory premises of competing paradigms, to what extent is it possible to sustain the claim that science is a fundamentally truth-seeking endeavor? In particular, Kuhn noted that scientific paradigms often incorporate foundational assumptions that are antithetical to the leading assumptions of succeeding paradigms, e.g., one cannot maintain an honest intellectual commitment to creationism and evolutionary theory without suffering from multiple personality disorder. It requires the intervention of an historical revisionist to invent a smooth, linear transition from one scientific paradigm to the next. As such, some critics have asserted that Kuhn's thesis exposed science as a fundamentally relativistic endeavor (Lakatos and Musgrave, 1970). In other words, the fact that successive paradigms tend to be epistemologically contradictory suggests that there is no essential consistency (i.e., no inherent truth) in scientific progress. That is, if scientific truth is linked to the assumptions upon which scientific paradigms are founded and, in turn, if scientific paradigms are disposable, then even in the most rigorous scientific endeavors truth must be only a provisional, transitory standard. In a world of paradigm shifts, truth would appear to be a chimera.

In the next chapter, I will examine the relationship between truth and science in greater detail. As contradictory as the foundational assumptions of competing paradigms may be, I will argue that, in contrast with Kuhn, scientific revolutions are in fact founded upon a consistent definition of truth. Indeed, I will argue that transitions from one paradigm to the next can be characterized as a process of redefining reality. In agreement with Karl Popper, I will emphasize that no single paradigm will ever be able to capture the universal, ultimate truth. Nevertheless, the process of redefining reality is predicated upon a cognitive renegotiation of the conceptual parameters of reality which is guided by an evolving standard of truth.

Chapter 2
Did Darwin Tell the Truth?

Charles Darwin is one of the most widely revered and enduringly controversial figures in the history of science. Both are exceptional feats for such a mild-mannered gentleman. Much of the controversy surrounding Darwin concerns the presumptive truthfulness of his evolutionary theory. Darwin's ideas about evolution were so ground-breaking that, more than one hundred and fifty years after the publication of *On the Origin of Species* (1859), many people still refuse to accept Darwin's basic precepts (McGettigan, 2010).

For all of the above reasons, Darwin and his work offer an illuminating illustration of the difficulties associated with defining the concept of truth. In some respects, truth might seem to be a transparently obvious concept: truth is what we see, taste, touch, smell, believe, etc. However, establishing general agreement about the scientific definition of truth has proven to be exceedingly difficult. Scholars have debated endlessly about whether truth should be defined as that which we observe vs. that which we believe, while others have argued that knowledge-seekers should abandon the search for truth altogether.

Again, Darwin is often credited with revealing some of the most profound truths ever discovered by a biologist, or, for that matter, by any scientist. On the other hand, Darwin's critics counter that his ideas represent a corruption of truth and are, thus, based upon a fundamental misperception of the facts (Behe, 2007; Dobbs, 2005). To evaluate these competing claims, I will draw upon a debate between Karl Popper and Thomas Kuhn, two scholars who held dramatically different views regarding the nature of science and the definition of truth. Though neither Popper nor Kuhn provide a sufficient rationale to finalize the debate over Darwin's ideas, both scholars help to explain why this particular controversy has raged so long. Finally, in spite of the fact that Kuhn and Popper do not offer the final word on role of truth in science, I will propose a remedy to this conundrum by contending that Darwin established the truth of biological evolution because his theory successfully redefined reality.

An Unlikely Revolutionary

While in college, Darwin entertained ideas of becoming a famous scholar. However, his ambitions were largely focused on supplementing the contributions of leading scholars in the field of natural philosophy. Early in his career, Darwin was a bright, but conventional thinker. Though he sought fame, at least initially, Darwin was not intent upon revolutionizing scientific conceptions of life as we know it. Indeed, if the truth be told, like many naturalists of his day, Darwin began his career as an earnest believer in creationism, i.e., the notion that life on earth was created in its existing form by an omnipotent Creator (Darwin, 1887, p. 39). Interestingly, Darwin's journey from credulous creationist to arch-evolutionist is arguably a microcosm of the larger intellectual movements that were shaking up the scientific landscape in the late 19th century.

Throughout his life, Darwin was an astute observer of the natural environment. Indeed, even in his youth, Darwin demonstrated a passion for collecting specimens that bordered on the fanatical. This is well illustrated by a story that Darwin recounted regarding his early experiences as an amateur field biologist (Darwin, 1990, p. 62). After peeling a piece of bark from a tree, Darwin spotted two varieties of beetles on which he had never clapped eyes before. No sooner had Darwin snatched up the bugs, than he witnessed a third new species caper into view. Having both hands occupied, and lacking any containers within easy reach, Darwin decided to free up his right hand by popping one of the captive bugs into his mouth. However, as soon as the insect landed on his tongue it ejected some type of horrible effluent. As a result, Darwin was forced to spit the beetle from his mouth. As he doubled over in a fit of gagging, both of his former captives as well as the new specimen scuttled to safety. A small, but significant loss to science.

Light-hearted as this story may seem, it is illustrative of the passion with which Darwin and other 19th century naturalists undertook their investigations of the natural environment. Beginning a century or so earlier, the Enlightenment had swept through Europe, shedding dazzling new intellectual light on a previously dark continent. Setting the stage for the Enlightenment, explorers, such as Christopher Columbus (Morison, 1942), had ventured beyond the pale and introduced their European brethren to mind-blowing new truths: new hemispheres, new oceans, new peoples. In many ways, Columbus did for the globe, what Galileo had done for the heavens. Each explorer had defied age-old beliefs, embarked on voyages of discovery, and produced compelling empirical evidence that the universe was larger, more complex and more full of mysteries than anyone had previously imagined.

Such discoveries were not always welcomed with open arms, nor did they immediately open minds. Just as Galileo had been rebuffed for posing a threat to medieval power structures, Columbus's discovery of the New World initially sparked disappointment, at least among his financiers (McNeese, 2006). Nevertheless, like it or not, Galileo and Columbus each opened the way to entirely new realms of discovery. Building on the crux of their achievements, Enlightenment thinkers constructed a real world philosophy that advocated an entirely new path to enlightenment (Hankins, 1985). At notable junctures during the Dark Ages, faith had trumped facts: contemplation of the cosmos had been restricted to those phenomena that complied with authorized views of creation. In such a context, "the enlightened" were defined as those whose faith sustained the intellectual scheme that gave order and meaning to the cosmos. During the Enlightenment, however, a new breed of philosophers argued that empirical exploration offered a more promising path to intellectual progress.

Certainly, there was an element of secular radicalism in the emergent Enlightenment philosophy. One could easily view Galileo's decades-long conflict with the Inquisition as the epicenter of a tectonic rift between science and Catholic theology (Feldhay, 1995). However, one must also keep in mind that Galileo remained a devout Catholic to the end of his days. Of course, Galileo may have maintained his faith for purely pragmatic reasons: to disavow Catholicism would have further inflamed the Inquisition's wrath. Nevertheless, Galileo and other path-breaking scientists were often people who possessed deep religious convictions (e.g., Newton, Linnaeus, Agassiz, etc.). Thus, it would not be accurate to characterize science as an enterprise that was intent upon destroying the foundations of theology. Instead, many early scientists embarked on their inquiries into the natural environment with the express purpose of seeking "god's perfection in nature" (Barrow, 2009, p. 19).

Although one might imagine that rigorous empirical observations, such as those produced by Galileo, would almost certainly generate enlightened thinking that deviated from the parameters of religious faith, in the 19th century, theologically-minded natural philosophers clung to a different view. According to Biblical Creation theology, an all-powerful and perfect God had created the universe in six days. As such, systematic study of the natural environment should, so the faithful believed, gradually yield evidence of God's sublime handiwork (Ray, 1691). If previous empirical studies had failed to generate evidence that supported the Christian view of Creation, then natural philosophers reasoned, the studies in question had simply failed to be sufficiently exacting. For example, Carolus Linnaeus once stated, "the closer we get to know the creatures around us, the clearer is the understanding we obtain of the chain of nature, and its harmony and system, according to which all things have been created" (Cited in Lindroth, 1983. p. 16).

Linnaeus' chief rival, Georges Cuvier, assembled exhaustive collections of animal species in order to catalog the extraordinary scope of God's creativity (Dobbs, 2005, p. 17). Interestingly, when Cuvier began studying fossils, he was struck by the fact that species which thrived in one geological epoch were often altogether absent in subsequent eras. Based upon this observation, Cuvier progressed to the startling conclusion that some species had gone extinct. For creationists, this was alarming idea because it seemed to imply that the Creator might be imperfect. That is, creations that exhibit undeniable imperfections (e.g., extinction) would also seem to imply a certain level of fallibility in their Creator (Barrow, 2009, p. 18-19). In addition, evidence of extinction also seemed to imply that Genesis might not represent a literal description of Creation. That is, according to Genesis, God created the universe and all of its inhabitants during one week-long burst of creativity. Thus, the life forms populating the earth were presumed to be the direct offspring of God's original creations. Humans were the children of Adam and Eve, tigers were the offspring of God's original tiger, bald eagles were the descendants of God's progenitor bald eagle, etc. Thus, evidence which suggested that the Earth had undergone repeated extinction events also undermined Genesis as a theory of Creation.

Natural philosophers acknowledged the dissonance between Cuvier's fossil evidence and the logic of Biblical Creation. However, instead of abandoning their perspective, theologically-minded natural philosophers, such as William Buckland (1824), simply rationalized extinction with other Biblical themes. Thus, Buckland argued that, whereas Genesis described Creation, the Bible also included examples of divine destruction: the Flood. For creationists, the Biblical deluge successfully explained why species that had flourished during one geologic era would suddenly vanish. Indeed, the fossil record revealed that when God chose to eradicate life, He tended to eliminate a great number of creatures with sudden, cataclysmic swiftness. Consequently, natural theologians developed a catastrophist school of thought that extended the creationist perspective to account for formerly anomalous mass extinctions (Grant and Estes, 2009, p. 36).

Thus, the marriage of natural philosophy and Christian theology. In their fact-finding endeavors, natural philosophers often stumbled upon evidence that cast a shadow on Biblical Creation. However, philosophical geniuses, such as Agassiz (1868), Buckland (1824), and Linnaeus (Blunt, 2004), rationalized anomalies in such a way as to better illuminate the truth, perfection, and beauty of God's divine plan. It was with such goals in mind that Darwin embarked on his career as a natural philosopher.

The Leisurely Radical

Though he remained an astute observer of the natural environment throughout his days, Darwin began as more of a hobbyist than a scientist. Having been born to privilege, the young Charles Darwin took full advantage of the perquisites afforded by his social position. Indeed, he grew so committed to a life of recreation that his father, Richard Darwin, nearly despaired. In an effort to steer Charles toward some type of productive career, in 1825, Richard sent his son to Edinburgh University. Though he was supposed to study medicine, Charles soon grew bored of the lectures. He was also repulsed by the horrors of surgery prior to the days of anesthesia. As a result, Charles neglected his studies. After two undistinguished years at Edinburgh, Richard decided that his prodigal son should transfer to Cambridge University where Charles could train to become an Anglican parson.

Lacking any sincere career passions, Darwin amiably complied with his father's wishes. He had no particular objection to life as a country parson. In fact, such a vocation might enable Charles to remain devoted to his life's greatest passions: collecting and recreation.

Unexpectedly, while at Cambridge, Darwin became captivated by some of the world's most eminent natural philosophers, several of whom would have a resonating impact on the balance of his life (Darwin, 1958). Perhaps most significantly, Darwin struck up a friendship with John Henslow, a leading figure in the field of natural theology—a branch of theology which asserts that profound religious revelation can be inspired by reason and experience. This, in combination with geological training from Alan Sedgewick, equipped Darwin with the necessary skills to take advantage of a history-making opportunity.

The Beagle's Naturalist

Shortly after graduating from Cambridge, Darwin received an invitation to ship out on the HMS Beagle. In addition to playing the role of messmate to the Beagle's Captain, Robert FitzRoy, Darwin was also invited to serve as the Beagle's amateur naturalist (Darwin, 1845). John Henslow had recommended Charles for the position because of his obvious enthusiasm, astute powers of observation, and solid academic training. Although Charles expressed unreserved enthusiasm for the venture, his father initially opposed the idea. From Richard Darwin's perspective, the proposed journey represented nothing more than another ploy by his son to postpone a life of professional responsibility. Fortunately, for Charles, his Uncle Josiah Wedgewood intervened and convinced Richard Darwin to support Charles' foray on the Beagle.

Darwin's journey on the Beagle stands out as one of the world's most important scientific events (Quammen, 2006). The journey was history-making not only because modern biology owes its existence to Darwin's circumnavigation, but, more specifically, because of the crucial change in Darwin's thinking that the voyage inspired. Darwin embarked on the HMS Beagle as an avowed creationist. Again, while at Cambridge, Darwin had consumed William Paley's *Natural Theology* without suffering a noteworthy case of intellectual indigestion (Grant and Estes, 2009). Paley's perspective on the natural environment supported many of the dominant themes in Christian theology. One such theme was the immutability of species.

At the time of Darwin's departure on the Beagle, creationists were openly hostile to the idea that life forms could undergo species-level modification (Sunderland, 1988). Certainly, individuals within any particular species demonstrated a wide range of morphological variation: siblings often exhibited marked variations in size, shape, weight, color, etc. However, having been explicitly designed by God, species were presumed to be immutable. Horses always had been horses, and the same was true for every other species. Creationists believed that life forms were incapable of breaking out of the molds into which they had been cast by God.

On December, 27, 1831, Darwin set off on the Beagle with a lifetime of experience as an amateur naturalist, a strong sense of wonderment, and an eagerness to make a valuable contribution to the field of natural philosophy (Desmond and Moore, 2009). Once at sea, Darwin devoted himself to his duties as the ship's naturalist. Having been invited aboard the Beagle in an unofficial capacity, essentially, as a gentleman's companion for Captain FitzRoy, Darwin did not earn the official title of ship's naturalist until well after the journey had begun. However, Darwin proved to be such an avid observer of the natural environment and keen collector of exotic plant and animal species that, by the time the Beagle reached South America, Darwin had cemented his role as the ship's official naturalist. Indeed, the enduring fame of the Beagle's legendary circumnavigation is due almost entirely to the zeal and cleverness of its young naturalist.

At an early stage of the voyage, Darwin reported that he was getting along famously with Captain FitzRoy (Darwin, 1845). In addition to sharing a common social class background, at the beginning of the voyage, Darwin and FitzRoy also shared a common view about the origins of life (Grant and Estes, 2009, p. 47). As result of their convergent perspectives, shortly after meeting FitzRoy, Darwin gushed to his sister that Captain FitzRoy was his "beau ideal of a captain" (Darwin, 1958, p. 49). However, as the journey progressed, the relationship between Darwin and FitzRoy cooled. Certainly, this disaffection could be partly explained by the length of the journey. Living aboard ocean-going vessels for extended periods of time can tax any relationship. However, more significantly, as the journey progressed Darwin's ideas about the origins of life began to diverge dramatically from FitzRoy's.

An important stimulant to Darwin's evolving scientific ideas, and one of the few texts that Darwin carried aboard the Beagle, was Charles Lyell's newly published *Principles of Geology* (1832). Interestingly, Lyell's book had been a gift from FitzRoy, who was apparently eager to equip his budding ship's naturalist with the very latest in scientific thinking. Still, Lyell's *Principles of Geology* was a curious choice for the Captain given his devoutly creationist beliefs. That is, in addition to incorporating a comprehensive survey of the most up-to-date geological observations, Lyell's *Principles* also advanced a radical "old earth" geological theory.

Darwin studied Lyell's *Principles* carefully. Its dramatic departure from creationism lay in Lyell's theory of uniformitarianism. In essence, this perspective asserted that the forces responsible for shaping the world's geological formations were, in fact, at work in the present just as they had been in the past. As such, Lyell's theory represented a stark contrast with Cuvier's catastrophist perspective. Catastrophism asserted that the earth was young and largely static with the exception of occasional cataclysmic outbursts of change. By contrast, Lyell's uniformitarianism asserted that the earth was very old and subject to ongoing, but very slight change. While such a theory may not seem terribly revolutionary, nevertheless, for young-earth creationists, the implications of uniformitarianism were mind-blowing.

In essence, Lyell argued that creation had not taken place at a singular moment in the recent past. Rather, creation was an ongoing process: the natural forces which were responsible for geological change were continuous, slow-paced and nearly invisible. The wind and rain that erode the earth generally do so at the level of dust motes. Fair enough, it takes naught but a glance outdoors to observe the gently erosive forces of wind and rain. However, what boggled the mind was Lyell's contention that it was none other than those minuscule forces that were responsible for sculpting the world's most spectacular geological features. And here we arrive at the truly revolutionary element of Lyell's theory. If such a thing were true, then uniformitarian change could only be accomplished over extraordinary expanses of time. How many years would it take for the buffeting winds and seasonal rains to carve the Grand Canyon? The answer: Eons upon eons, and many magnitudes more years than could easily fit within the creationists' young earth paradigm.

Yet, impossible as it may have seemed, Lyell's theory made sense. If afforded sufficient time, the relentless forces of accretion and erosion could build peaks that touched the clouds and then, particle by particle, reduce the ruggedest range of mountains to a chain of low, rolling hills. As the Beagle cruised along the exotic coasts of South America, Darwin gazed upon Patagonia's sublime geology and considered the possibilities. The wonders of South America were almost too numerous to comprehend. While trekking through the Andes, Darwin found layers of fossilized shellfish that stood thousands of feet above sea level. In what distant age had those shellfish thrived in their millions upon the ocean's floor? And how many eons had it taken to thrust those long dead shellfish to the very tops of the Andes? It boggled the mind.

Even as Darwin gazed upon these Andean wonders, his mind strayed to the many specimens that he had collected during his journey, and, indeed, throughout his entire life. Had all those creatures been created in a single moment by God, or had some other force sculpted the endless varieties of beetles, bugs and birds that Darwin treasured?

Islands of Wonder

As the journey progressed, Darwin finally arrived at the Galapagos Islands (De Roy, 1839). Though he had witnessed many wonders during his travels, the bizarre menageries that he encountered in the Galapagos exceeded anything that he had yet imagined. From giant tortoises to endless varieties of land crabs and snails, Darwin marveled at the seeming adaptability and (dare he think it?) *mutability* of the species that he observed. Perhaps as he gazed upon the spectacle of marine iguanas bobbing in the surf, Darwin gave thought to a new and unsettling idea. Just as tiny and slow-paced geological changes had the net result, over the long haul, of introducing extraordinary alterations to the earth's geology, might not the same be true for living organisms? In other words, could the tiniest physiological changes accumulate sufficiently across time to bring about the transmutation of species?

Darwin was both fascinated and disturbed by the implications of this idea. Though other early scientists, including Darwin's grandfather, Erasmus Darwin (1796), had toyed with the idea that life had evolved through random natural processes, no one had any idea how that could happen. However, it was in the Galapagos that Darwin finally witnessed the diversity of life forms that would help him reveal the basic mechanisms of evolution.

In particular, while touring the Galapagos, Darwin noted a phenomenon that tends to arise with regularity among island species. After migrating to islands, species often undergo remarkable alterations, e.g., small animals grow, large animals shrink, dietary and habitat preferences shift, etc. Precisely why such a phenomenon should be so prevalent in diffuse island ecologies remained a mystery until Darwin studied the many varieties of finches on the Galapagos (Keynes, 2003). Galapagos finches exhibit such a wide variety of shapes and sizes that Darwin initially misclassified the birds as entirely different species. Indeed, it was only after consulting with John Gould, an expert in bird physiology, that Darwin realized that the finches were in fact much more closely related (Nelson, 2009, p. 289). As a result, Darwin experienced a revelation.

Darwin postulated that island migrants encounter unique population pressures. That is, when migrant species initially arrive on the shores of hospitable islands their populations tend to explode. However, success quickly becomes a migrant's worst enemy because rapid population growth tends to exhaust available resources. By the way, Thomas Malthus (2003) also had a substantial impact on Darwin's evolutionary thinking. Quite simply, Malthus argued that population tends to grow geometrically whereas food resources can only expand arithmetically. As a result, if unchecked, population growth among any successful species (i.e., lilies on a pond, humans in Manhattan, finches in the Galapagos, etc.), will, within the space of only a few generations, rapidly exhaust available food resources. Under such circumstances, species that depend upon the seemingly limitless bounty of their local environs will soon discover that their luck has run out.

Thus, Darwin speculated that, like many island success stories, a singular species of finch had migrated from South America to the Galapagos. Finding itself in an environment that was largely devoid of predators and competitors, the finches flourished. However, like so many of their migratory counterparts, the finches soon encountered a problem. Finch populations expanded to the point that food became scarce and, thus, competition for the islands' limited resources became increasingly intense. In turn, the combined pressures of overpopulation, scarcity of resources, and the resultant competition for survival triggered a process that transformed the fortunes of finches on the Galapagos. It was this insight that enabled Darwin to crack the mystery of natural selection.

In brief, Darwin argued that there are a number of crucial biological dynamics that energize the evolutionary process:

1. Variation: whether it's dogs, grass, or fruit flies, organisms tend to vary from one individual to the next
2. Overpopulation: from oak trees to salmon, parents tend to produce more progeny than can survive to maturity
3. Struggle for survival: the overproduction of progeny tends to inspire a high-stakes competition to secure limited resources
4. Survival of the fittest: individuals with advantageous genetic traits enjoy an edge in the competition for scarce resources
5. Evolution through natural selection: winners of bio-ecological competitions survive and pass advantageous genetic traits to their offspring—which, in turn, brings about the gradual transmutation of species

With the above in mind, Darwin asserted that the finches which had survived on the Galapagos were those that had developed some sort of competitive advantage over their counterparts. As a result, multiple subspecies emerged among Galapagos finches, each of which was genetically equipped to take advantage of resources that were distinct from those preferred by their former messmates (De Roy, 2009, p. 122-129). Thus, Galapagos finches may have arrived on the islands as a single species, however, due to survival pressures, the finches evolved into a wide array of subspecies with distinct body types, divergent diets and unique survival strategies. Furthermore, Darwin argued that the speciation process that had transmuted Galapagos finches was essentially the same for every other species. Therefore, Darwin concluded that every living creature owed its existence to a process of evolutionary transmutation (Darwin, 1859).

A Scientific Bombshell

As one might imagine, Darwin's ideas represented a huge and, for many, vexing transformation in scientific thinking. Whereas Cuvier had stretched creationist thinking to explain fossil evidence of extinction, there was simply no way to incorporate evolutionary theory into the creationist perspective. As Thomas Kuhn (1962) would later put it, creationism and evolutionary theory are distinct and incommensurable paradigms. Darwinian evolution attacked every one of the foundational assumptions upon which creationism was based. Whereas creationism suggested that the earth was relatively young, Darwin, in agreement with Lyell's old earth geological paradigm, argued that the earth was inconceivably ancient. In addition, whereas the creationistic perspective asserted that species were immutable (i.e., all species had been intentionally designed at one point in time by an all-knowing, perfect creator), Darwin argued that species routinely transmuted as a result of random evolutionary pressures. As a result, evolution did more than shed new light on the processes through which life had evolved on planet earth, evolution also proposed an entirely new secular cosmology. Although Darwin's evolutionary perspective did not rule out the possibility that God might exist, nevertheless, evolution made God irrelevant in the origin of species.

Thus, Darwin's evolutionary theory offered not just a new and exceptionally far-reaching scientific perspective, evolution also represented an unforgivable affront to anyone with an affinity for creationism. In Christian theology, God was not only the source of Creation, but God was also the most sacred being in the universe. Though one could argue that Darwin was just trying to be a good scientist, from the moment that Darwin published *On the Origin of Species*, his theory has been attacked for being a slur upon the sanctity of creation and, worse, as an insult to God.

Being keenly aware of the firestorm that his ideas were likely to instigate, Darwin delayed publication of his theory. Although Darwin shared early versions of *On the Origin of Species* with friends and colleagues, he procrastinated for an inordinately long time. Indeed, Darwin dawdled for so long that he almost failed to receive due credit for formulating his evolutionary theory. In 1858, Alfred Russell Wallace posted a letter to Darwin from the Malay peninsula in which he proposed a theory of natural selection that, though derived independently, had strong similarities to Darwin's. Although Darwin could have ignored the letter and claimed sole credit for the theory of natural selection, being a conscientious scholar, Darwin invited Wallace to co-present a paper before the Royal Society. Russell accepted and, as a result, to this day, credit for originating the theory of evolution has been shared (Darwin and Wallace, 1858).

Honorable as his actions toward Wallace may have been, Darwin has been roundly criticized for the sluggish pace at which he published major scientific findings. Some have argued that, being uncomfortable with controversy, Darwin may have delayed simply because he did not wish to incite public antipathy. Perhaps even worse than public censure, Darwin might have been concerned about the effect that his theories would have on his personal relationships (Keynes, 2002; Quammen, 2006). Many people who were near and dear to Darwin—not least of which was his wife—were devoutly religious and, thus, Darwin may have feared that his ideas would damage those relationships.

Apart from his personal concerns, one could also argue that Darwin delayed publication for scientific reasons. Being cognizant of the outrage that his theory was certain to inspire, Darwin may have wished to ensure that his theory was impregnable before revealing it to a censurious public. In short, Darwin may have delayed simply because he was a good scientist.

Getting it Right

Whether Darwin temporized because he was fearful of public opprobrium, or because he was one of the world's most exacting scientists is a matter of ongoing debate. Nonetheless, what is clear is that Darwin's evolutionary theory has inspired every bit of the controversy that Darwin anticipated (Smout, 1998). Indeed the scope and intensity of the controversy over evolutionary theory is sometimes difficult to comprehend (McGettigan, 2010). Rarely do scientific theories elicit even the slightest attention from the general public. Apart from a few celebrated scientists, such as Einstein, the workaday world of science usually operates off the radar screen of public interest. Not so with Darwin. Evolutionary theory has inspired widespread acrimony from the moment of its first publication (National Academy of Science, 2008).

In addition to other objections, many people have been displeased with the idea that, within the confines of evolutionary theory, humans do not occupy any special pride of place. From Darwin's perspective, humans were, quite simply, just another form of life (Darwin, 1871). Essentially, evolution postulates that humans exist because their ancestors randomly developed advantageous genetic traits. For those who are in need of a loftier sense of philosophical purpose, they won't get it from Darwin. Evolutionary theory remains profoundly agnostic concerning the philosophical meaning of life.

So radical was Darwin's new theory that, at first, even leading members of the scientific community expressed skepticism. For example, initially, Charles Lyell was only willing to accept evolution as an explanation for the transmutation of "lower animals," but not for humans (Grant and Estes, 2009, p. 238). Nevertheless, evolutionary theory gradually won over the majority of the scientific community. Scientists were swayed by evolution in part because, as a scholar, Darwin was rigorous to a fault. Having once drawn premature conclusions about the distribution of ice age rocks—and then suffering the indignity of being publicly upstaged by Louis Agassiz—Darwin determined never to make the same mistake again (Dobbs, 2005, p. 33-36). Throughout the balance of his career, Darwin made lengthy and exacting empirical observations and, only then, constructed theories that corresponded closely with the facts. Thus, Darwin's fact-driven approach to theory construction represented a huge advance for the fledgling biological sciences. Whereas creationism encouraged observers to interrogate facts through the lens of Christian dogma, Darwin set aside creationist preconceptions and then constructed his theory in such a way as to explain previously anomalous facts. As a result, Darwin has won the enduring admiration of his colleagues for being an exemplary good scientist.

Repaving the Landscape of Truth

Yet, from Thomas Kuhn's perspective, Darwin's enduring importance is primarily due to his invention of an exceptionally powerful paradigm. Evolution is capable of explaining the entire history of life on earth. In addition to unifying the biological sciences, evolutionary theory has also created bridges between scientific disciplines. Again, Darwin not only benefited from from Lyell's geological insights, but, thanks to Darwin, geologists now enjoy a more nuanced understanding of fossil-laden stratigraphy. Thus, by building conceptual linkages across scientific disciplines, Darwin's theory won additional support from scientists. Indeed, the degree to which evolutionary theory continues to enjoy broad interdisciplinary applicability (i.e, all the way from Anthropology to Zoology) explains the near uniform support for evolution among scientists.

Interestingly, although many scientists would argue that evolution represents a bedrock scientific truth (Dawkins, 2009), Kuhn would counter that evolutionary theory is merely a mature paradigm that has successfully marginalized competing perspectives. Again, for Kuhn, sophisticated as paradigms may be, given that they are the products of human intellectual endeavors, all paradigms are necessarily as imperfect as their creators. More crucially, Kuhn argues that the truthfulness of any particular paradigm is simply a measure of the dominance that the paradigm enjoys.

While Kuhn believed that Darwin's ideas represented an important paradigmatic alternative to the creationist perspective, he would not have said that Darwin's ideas were more truthful than creationist beliefs. Evolution was simply different. For Kuhn, asserting that scientific knowledge was evolving toward a predetermined ideal of truth was roughly equivalent to asserting that life forms were evolving toward a predestined ideal of perfection, e.g., Hitler's Aryan race. Kuhn argued that such prejudgments were undesirable. That is, Kuhn believed that imposing teleological outcomes on the process of scientific deliberation proceeds from the assumption that some privileged observer has the necessary foresight to understand and, thus, delineate optimal outcomes. Not only do such teleological theories tend to endorse one type of irrational prejudice or another, but teleological theorizing also legitimizes anti-scientific intellectual practices, i.e., privileging evidence that supports a preferred perspective, while prejudicially ignoring, downplaying or otherwise marginalizing countervailing evidence.

Kuhn believed that prescribing a singular intellectual destination (e.g., Truth) *before* scientists have even initiated their research has a tendency to bias the outcomes of scientific inquiry. In other words, the more faithful they are to a single set of orienting assumptions, the more inclined scientists might be to seek "truths" that happen to support their preferred paradigms. Thus, in an effort to liberate science from the malign abuses of paradigmatic bias, Kuhn proposed that scientists should discontinue their singular pursuit of truth—if only so that, during the course of their researches, scientists might be inclined to learn things that they were not singularly determined to discover from the outset.

Competing Perspectives

Whereas Kuhn believed that the pursuit of truth represented an illegitimate form of teleological intellectual endeavor, Karl Popper begged to differ. Popper (1959) was inclined to agree with Kuhn about the hazards of teleological theorizing. In particular, Popper railed against Marxists who believed that Marxist theory equipped its followers with a privileged perspective from which to judge the entire course of human history (Popper, 1950). Nevertheless, Popper believed that scientists needed to coordinate their knowledge-seeking endeavors around a clearly specified definition of truth, or else science would quickly devolve into an aimlessly relativistic pursuit.

Thus, for Popper, truth was a standard that made it possible to evaluate the relative veracity of competing knowledge claims. In contrast with Kuhn, Popper believed that some statements were true while others were demonstrably false. Truth was an objective standard that made it possible to differentiate between those two extremes. On the other hand, when knowledge seekers explicitly disavow the legitimacy of truth standards—as Kuhn advocated—then, Popper believed, it becomes impossible to distinguish in any absolute sense between truth and falsity. If there are no standards with which to affirm truth, then neither is it possible to castigate falsity. In a relativistic epistemological framework, knowledge is merely knowledge; neither true nor false, neither fact nor fiction. Relativism implies that all knowledge claims are equally (in)valid.

Given his realist philosophical perspective, Popper defined truth as knowledge that corresponds with the facts (Popper, 2002, p. 112). In this context, facts can be understood as empirically derived evidence, e.g., fossilized animals, feeding habits of Galapagos finches, erosion patterns on rocks, starlight measurably bending around the sun, etc. In addition, Popper argued that science was distinguishable from other intellectual endeavors because, quite simply, scientific knowledge has demonstrable truth content. To establish its truth content, Popper argued that scientific knowledge needs to be stated in such a way that its truth or falsity can be tested with reference to empirical evidence. Thus, proceeding from these assumptions, truth may be understood as knowledge that, when subjected to empirical testing, is found to be supported by the facts, whereas false statements can be understood as knowledge claims that are contradicted by the facts. Take, for example, the statement: "all species are immutable." In this instance, multiple scientific disciplines have generated an abundance of fossil, biochemical, experimental, geological, and genetic evidence which illustrates that species are in fact mutable. Therefore, given that the preponderance of empirical evidence contradicts the suggestion that species are immutable, it is possible to conclude that the statement, "all species are immutable," is false.

Having established an operational definition of truth, Popper argued that science would be more apt to progress through the process of falsifying untruths, rather than verifying truths. Citing Hume's philosophical dilemma, Popper pointed out that it was nigh on impossible to inductively verify that any statement was positively true. According to Popper, the "problem of induction" (Popper, 1959, p. 11) is such that one must establish the unvarying certainty of any statement in all times and places before being able to verify its absolute truthfulness. For example, to establish the truth that "all fish live in water," one would literally have to examine every fish at all times and places and verify that each and every one lives in water. Quite apart from the utter impracticality of generating a universe of positive support for every individual knowledge claim, Popper also pointed out that even the most seemingly truthful statements can be vetoed by a single negative example. As in the case of fish presumably living in water, there are numerous examples of flying fish and mud-dwelling fish which would be sufficient to dispute that particular claim. In addition, one must also consider the dimension of time. Although the vast majority of fish currently make their homes in aquatic environments, many terrestrial animals owe their existence to fish that, long ago, hauled themselves out of their watery habitats and adapted to a terrestrial lifestyle. Thus, Popper concluded that scientists could never hope to establish the truth of any statement through an inductive analysis of positively supporting evidence.

Just as it is impossible to verify truths inductively, Popper also noted the fundamental flaws in deductive reasoning. Whereas induction proceeds logically from the particular to the general, deductive logic proceeds from the general to the particular. As such, Popper (1950) critiqued a variety of deductive rationales (e.g., Marxism, Freudian and Adlerian psychology, etc.) for committing the anti-scientific error of privileging theory over evidence. In other words, deductive reasoning tends to emphasize, first and foremost, faith in a particular theoretical perspective. As such, in practice, deductive logic has a tendency to privilege evidence which supports the assumptions of one's preferred orienting theory while discounting, suppressing or otherwise failing to properly consider countervailing evidence. For example, no matter how well evolution may explain the relevant scientific facts, from the creationist standpoint evolutionary theory appears wrongheaded—precisely because evolution puts facts before theory. All too often, facts tend to be judged, or deduced relative to one's beliefs. For example, if someone believes that God created the universe in six days, then it is simply a matter of faith to deduce that God also sculpted the Grand Canyon, complete with all of its fossil-laden stratigraphy, in naught but the blink of an eye. For creationists, the universe exists as God made it and for reasons that are scrutable only to God (Ackerman, 1986; Collins, 2006). No further explanation is required. From a creationist standpoint, the relevant facts are deductively compelling and, thus, verify rather than falsify a continuing belief in creationism. As in any debate, depending upon one's orienting assumptions, the interpretation and perceived importance of seemingly objective facts can conflict diametrically.

In a different context, Kuhn (1962) developed a similar critique of deductive reasoning. From Kuhn's perspective, supporters of various paradigms have a tendency to gauge the relative merits of empirical facts through the lens of their preferred paradigms. For example, instead of viewing Galileo's novel observations as a spectacular achievement in the realm of scientific intellectual endeavor, advocates of geocentric theory elected to attack Galileo and suppress the evidence that inspired more rational minds to acknowledge the merits of heliocentrism.

Falsification

Popper's breakthrough was to establish the principle of falsifiability (Popper, 1959, 1965). This principle asserts that testable knowledge claims can be considered valid until they are effectively disproved, or falsified. In other words, Popper argued that all knowledge was provisional. That is, objective, universal truths might exist (Popper, 1972), but they can never be adequately summed up by human knowledge claims. While such an epistemological position might seem counterproductive (i.e., truth claims can never be positively verified, rather they can only be conclusively refuted), Popper's falsifiability principle actually articulates a working compromise between the pitfalls of inductive and deductive logic.

To begin with, Popper embraces the goal of developing increasingly truthful knowledge through a process of ongoing problem solving (Popper, 1999). While rejecting a postivistic approach to inductive science, Popper emphasizes that induction is essential to the process of generating increasingly verisimilar scientific knowledge. Essentially, Popper argued that scientific knowledge has a tendency to advance incrementally through the process of conjecture and refutation (Popper, 1965). Conjectures, from Popper's perspective, are testable statements that posit some sort of original insights about the nature of the empirical universe: for example, Einstein's argument that space is curved. Popper was impressed by this conjecture because it represented a highly innovative view of the empirical universe and, secondly, because it was testable. In point of fact, Arthur Eddington (Wolfson, 2003, p. 211-212) conducted a famous test of Einstein's conjecture by traveling half way around the world to measure the flow of starlight during a solar eclipse. That is, instead of traveling in straight lines, if space was indeed curved, then light should bend around massive objects such as the sun—and this phenomenon should be especially observable during a solar eclipse. Ultimately, as a result of his measurements, Eddington produced support for Einstein's theory of relativity.

While Eddington's results were widely celebrated as a resounding confirmation of Einstein's theory, Popper emphasized that, in this and every other case, scientific progress is better served via the process of systematic refutation, rather than verification. Not only do refutations of popular theories suggest that there is sufficient openness in a particular intellectual environment to foster free thought and inquiry (Popper, 1950), but refutations also tend to instigate the production of newer, better conjectures. Popper believed that, in science, no idea should be sacred, nor any personage be too venerable to critique (Popper, 1982b).

Powerful as Einstein's theories may have been, even Einstein was prepared to admit that his conjectures were far removed from the ultimate laws that governed the universe. Indeed, Einstein associated his concept of God with the infinite mysteries that make up the universe. Regarding the extent to which modern science had been able to compass those infinite mysteries, Einstein had this to say:

> We are in the position of a little child entering a huge library filled with books in many languages. The child knows someone must have written those books. It does not know how. It does not understand the languages in which they are written. The child dimly suspects a mysterious order in the arrangement of the books but doesn't know what it is. That, it seems to me, is the attitude of even the most intelligent human being toward God. We see the universe marvelously arranged and obeying certain laws but only dimly understand these laws (Quoted in Isaacson, 2007, p. 386).

Though he respected Einstein deeply (Popper, 1982b), Popper would have been as eager to see Einstein's theories refuted as he would the theories of any other scientist. For Popper, science was an intellectual journey. The goal of scientific inquiry was to sustain evolutionary intellectual progress (Popper, 1972). For Popper, scientific progress amounted to an iterative problem-solving process. Via the cycle of conjecture and refutation, science incrementally progressed from one intellectual evolutionary stage to the next—always oriented toward, but never fully achieving the final goal of producing universal truths. As such, just as the incremental accumulation of seemingly minor mutations can gradually bring about the emergence of an entirely new species, Popper asserted the that gradual, step-by-step conjecture and refutation tends to produce entirely new "species" of intellectual endeavor.

Evolution is in the Eye of the Beholder

Thomas Kuhn viewed the evolution of scientific ideas in a much different light. As opposed to stepwise process of scientific conjecture and refutation, Kuhn argued that scientific revolutions take place through seismic dislocations in the history of ideas. As evidence, Kuhn points out the striking contradictions that are often evident in the foundational assumptions of divergent paradigms (e.g., geocentrism vs. heliocentrism, creationism vs. Darwinian evolution, classical mechanics vs. relativity, etc.). Rather than being suggestive of incremental progress, Kuhn argues that scientific revolutions are better understood as hostile intellectual takeovers. There tends to be little love lost among the adherents of divergent paradigms. Indeed, the intellectual differences between scholars of varying paradigmatic inclinations tend to be so pronounced that, rather than compromising, paradigm shifts usually take place only when avowed supporters of once dominant paradigms gradually retire and/or expire. That being the case, if it is appropriate to speak of evolution with respect to the history of ideas, then Kuhn's version is catastrophist, whereas Popper's is uniformitarian.

Interestingly, both Kuhn and Popper's descriptions of scientific evolution are applicable to the scientific enterprise, but under markedly different circumstances. As Kuhn was willing to concede, scientific evolution is marked by extended periods of time during which dominant paradigms enjoy broad support in a particular discipline. During such relatively quiescent time frames, which Kuhn referred to as periods of normal science (Kuhn, 1996, p. 10), scientists work collaboratively to help science progress in a stepwise fashion. Essentially, scientists that operate within the same paradigmatic perspective tend to view facts through the same conceptual lens and, as such, they can achieve broad agreement about the truths that facts seem to reveal. For example, the field of biology is currently unified around the theory of evolution; though biologists study many different aspects of life, they all speak the common language of evolutionary theory. In other words, during periods of normal science, paradigms are sufficiently dominant that the majority of scientists are able to frame their various research programs around similar assumptions, e.g., all life evolved from a common ancestor. As a result, during periods of normal science, it is possible for Karl Popper's uniformitarian version of scientific progress to dominate.

Lively as normal scientific debates can be, participants usually focus on details as opposed to repudiating the foundational assumptions of entire scientific fields. For example, the Homo floresiensis find has stirred passionate debate in the field of anthropology (Henneberg and Schofield, 2008). However, arguments tend to focus on precisely where hobbit-like humans ought to fit in the grand scheme of anthropoid evolution. Disagreements over Homo floresiensis generally do not call into question the fundamental veracity of anthropological science itself. In other words, although disagreement is endemic among knowledge-seekers, the scientists who are engaged in normal scientific debates still speak the same language and conduct their debates within the parameters of conventional scientific discourse. That is, the introduction of astonishing new facts (e.g., hobbit-sized humans), encourages scientists to develop new conjectures that expand the boundaries of scientific inquiry in a way that permits previously unknown facts to be integrated into the established body of scientific knowledge. Thus, Popper's stepwise accumulation of knowledge.

On the other hand, it is possible for scientific debate regarding the nature of fossil evidence to diverge from the normal path and instead pursue a more revolutionary intellectual agenda. Once again, upon encountering facts (i.e., fossilized evidence of extinction) that did not readily fit within the creationist paradigm, Georges Cuvier and other scholars developed a catastrophist perspective that remedied a number of shortcomings in the creationist perspective. Through the more sophisticated lens of catastrophism, fossilized evidence of extinction revealed new truths about the complex process of divine creation and destruction. Yet, when Darwin examined very similar fossil evidence, he concluded that essentially the same facts revealed entirely different truths.

Again, Darwin began his career as a credulous disciple of creationism. Consequently, it would have made sense for Darwin to assess facts, cling to established truths, and normalize anomalies in a fashion very similar to Georges Cuvier. Yet, rather than creatively reinventing his guiding paradigm and, thus, preserving and validating the truths that buttressed creationism, Darwin struck out in an altogether new intellectual direction: Darwin posited a biological, rather than a divine explanation for the origin and extinction of species.

Impressive as Darwin's breakthrough may have been, it was also somewhat puzzling. If, as Popper argues, the truth can be defined as "knowledge that corresponds with the facts" (Popper, 1972, p. 323), then one must wonder how the very same facts could inspire such divergent interpretations of reality. In other words, if truth is supposed to be a type of knowledge that is consistent at all times and places, and, if truth is knowledge that corresponds with the facts, then truthful evaluations of the facts should generate fairly consistent interpretations of empirical reality. In other words, if facts reveal truths, then Darwin and his creationist counterparts should have arrived at essentially the same conclusions about the origins of life. That Darwin and creationists could derive such emphatically divergent interpretations of bedrock facts would seem to undermine Popper's contention that facts correspond with the truth. Instead, the discontinuities between evolution and creationism seem to support Kuhn's assertion that truth is an entirely context-dependent knowledge standard, and new paradigms emerge from unequivocally incommensurable scientific advancements.

Revolutionary Facts

Instead of expanding scientific frontiers through a stepwise problem-solving process, Kuhn argued that the primary instigator of scientific progress was intellectual discontinuity. Kuhn argued that paradigms come and go because new facts invariably emerge that defy explanation within existing paradigms. For example, the young-earth creationist perspective cannot easily account for ancient marine fossils that are situated atop mountain peaks.

Kuhn referred to such inexplicable facts as anomalies. Paradigms that fail to account for anomalies tend to inspire more doubt than confidence. As the number and variety of anomalies increase, Kuhn argued, paradigms often enter a stage of crisis (Kuhn, 1996, p. 77). During paradigm crisis, admirers often go out of their way to preserve the influence of a weakening paradigm, e.g., ignoring or rationalizing facts that threaten the explanatory power of the paradigm. On the other hand, dissenters take advantage of paradigm crisis to advance competing paradigms. Generally speaking, newly formulated paradigms are designed to explain all of the anomalies that preexisting paradigms have failed to account for. Thus, Darwin's theory of evolution provides ready explanations for anomalies that remain enigmatic to the creationist perspective, e.g., evidence of the mutability of species, evidence of the earth's antiquity, fossils indicating evolution from common ancestors, etc.

Thus, Kuhn acknowledged that emergent paradigms generally have a greater explanatory power than the paradigms that they replace. Yet, in spite of their augmented explanatory capabilities, Kuhn was unwilling to concede that new-

and-improved paradigms have greater truth content than the antiquated paradigms that they replace.

> Newton's mechanics improves on Aristotle's and . . . Einstein's improves on Newton's as instruments for puzzle-solving. But I can see in their succession no coherent direction of ontological development. On the contrary, in some important respects, though by no means in all, Einstein's general theory of relativity is closer to Aristotle's than . . . to Newton's (Kuhn, 1996, p. 31).

Though Kuhn acknowledged that there were often demonstrable differences between competing paradigms, Kuhn did not believe that the merits of various paradigms were discernible with respect to an objective standard of truth. If anything, Kuhn argued, the ephemeral nature of paradigms as well as the manner in which divergent paradigms tended to espouse diametrically opposed truths emphasized that the evolution of scientific knowledge progressed independently of a consistent standard of truth.

Wherefore Truth?

Therefore, in answer to the question, "Did Darwin tell the truth?," Thomas Kuhn would have replied, "No." Since Kuhn believed that all knowledge was relative, he admitted to no objective means by which to determine the superiority of evolutionary vs. creationist truths. Consequently, Kuhn would likely have contended that, if supporters of evolutionary theory believed that their paradigm was an exemplar of the truth, then they were at liberty to cling to that belief—as, for that matter, were supporters of creationism.

Kuhn's argument has enjoyed an enduring appeal. Postmodernists employed Kuhn's thesis to advance a thoroughgoing critique of science. The postmodern critique was also founded upon the basic notion that truth was the central myth of modern science. Of course, the problem with this critique is that, in the absence of truth, facts can be interpreted in any fashion that observers desire. Again, Kuhn argues that, even in situations where observers hold diametrically opposed opinions about facts (e.g., species are mutable vs. immutable), there is no way to decide which observation is more truthful. Again, from Kuhn's perspective, truth is a function of the paradigmatic priorities that one embraces. If one's preferred paradigm emphasizes that species are immutable, then one's convictions to immutability represent as sound a statement of truth as any other.

Perhaps he was a spoilsport, but such conclusions did not sit well with Karl Popper. Even though it might be possible to interpret facts in a variety of ways, Popper was convinced that some interpretations were true while others were false (Popper, 1983). Popper believed that the best way to distinguish between those extremes was through a careful, systematic interrogation of the facts (Popper, 1972). Of course, the problem with such a perspective is that, as Kuhn pointed out, paradigmatic priorities tend to predispose an observer's frame of reference: absolute truths in one paradigm (i.e., mutable species, or curved space) tend to be viewed as egregious untruths from other perspectives. Thus, it would seem that the only way to generate unbiased interpretations of facts would be in a context that is devoid of paradigmatic prejudice. But how could such a thing be possible? Under what circumstances can observers wrestle free from the distorting biases of paradigms?

Certainly, Kuhn argued that, in almost every case, individuals were swayed by the influences of one paradigm or another. There was, however, one exception to this general rule: paradigm revolution.

Again, Kuhn argues that knowledge-seekers are often confronted with anomalies, or facts for which existing paradigms lack adequate explanations. In response, observers can elect to respond in one of three ways:

1. Deny the relevance of anomalies (e.g., creationists who disregard evidence of the mutability of species)
2. Attempt to shoehorn the anomalies into a preferred paradigm (e.g., as Cuvier accomplished with his theory of catastrophism)
3. Develop entirely new explanations for the anomalies

In agreement with Kuhn, Popper argued that, when confronted with inexplicable facts, observers will often develop innovative conjectures (Popper, 1965) that are intended to render anomalies intelligible (e.g., Einstein's explanation of Brownian motion, Alfred Wegener's theory of continental drift, or Darwin's notion of radial evolution). Popper contends that generating new conjectures is predicated upon an elemental wellspring of creativity (Popper, 1988, p. 122-123). Although Popper conceded that the source of such miraculous creativity may be unknowable, he did emphasize that this enigmatic fount of human ingenuity was an essential source of inspiration for new conjectures. Thus, both Kuhn and Popper argue that individuals possess the ability to invent original ideas that often contradict the assertions of established paradigms. On a variety of occasions, I have referred to this capacity as agency (McGettigan, 1998, 1999, 2002, 2006). Agency can be understood as an individual-level capacity to conceive novel ideas and, importantly, modify the existing structure of knowledge.

Both Popper and Kuhn argue that encounters with empirical facts and, in particular, anomalies, help to stimulate the production of novel scientific explanations. What each failed to realize, however, was that the process of developing entirely novel scientific explanations is a creative process that is guided by and is productive of truth. I refer to this process as redefining reality (McGettigan, 1998, 1999, 2002).

Redefining Reality

Redefining reality is a process through which individuals can challenge inadequate paradigms through a combination of astute observation and an ingenious capacity for innovative cognition (i.e., agency). The notion of redefinable reality posits, in agreement with Popper's realist philosophy, that there is a universe "out there" that exists independently of human cognition (Popper, 1983). As such, I argue that universal Truth does exist, but such Truth is not (nor will it ever be) contained within extant scientific paradigms. Rather, The Truth extends infinitely into the unlocked mysteries of the expanding universe. In other words, reality is what it is: an asteroid is an asteroid is an asteroid, etc... Truth is an intrinsic, inseparable feature of phenomena as they exist independently of human perception. Lies and distortions come into existence via humanity's vast capacity for ignorance: humans view the illimitable universe through awed and flawed psyches. Although admirable in many ways, the human grasp of infinite mysteries remains woefully incomplete. Nevertheless, the process of redefining reality permits limited human psyches to transcend the limitations of inadequate paradigms in pursuit of a grander vision of Truth.

The process of redefining reality often begins when agents make unanticipated observations, e.g., "Hey! Galapagos finches are more closely related than I thought." Individuals may follow up such observations by issuing a challenge to established paradigmatic restrictions, i.e., "This seems to suggest that, rather than being immutable, island species undergo transmutation." In the process of attempting to make sense of such anomalies, individuals tend to deconstruct (Derrida, 1978) the conceptual frameworks that limit their ability to comprehend mysterious phenomena, i.e., "Based upon what I have observed, I no longer believe that species were created in their present form by God." As individuals reevaluate their beliefs with respect to their inability to comprehend anomalies, the features of their paradigms that do not hold up under scrutiny come under substantial erosive pressure. If individuals are persistent enough, they may reach a point at which the critical mass of their contemplations overloads the shackles of their former beliefs and, thus, they may experience a moment of truth, i.e., "Aha! Species evolve as a result of natural selection."

Moments of truth are similar to eureka experiences wherein, having deconstructed the distorting influences of inadequate paradigms, individuals successfully invent a more satisfactory definition of reality. These experiences may be considered relatively truthful in that they are generated through a process that involves the intentional negation of social controls over an individual's definition of reality. This is not to say that the redefined paradigm at which one arrives after experiencing a moment of truth is, therefore, Truth. Far from that, in keeping with the assertions of radical power theorists (Foucault, 1977; Lukes, 2005), I maintain that all established belief systems exert their own forms of ideological power on the construction of knowledge. Thus, to experience a moment of truth does not transport one to an ideal realm wherein Truth reigns unchallenged—as opposed to the assertions of Habermas (1970, 1972, 1981). Instead, I merely suggest that the process of redefining reality permits individual agents to experience moments of truth within the ideologically-coercive domain of social reality. With the help of such redefined insights, agents become better equipped to negotiate with the pervasive, consciousness-distorting influences of radical power sufficiently to transcend the limitations of established paradigms for the purposes of creating better (but never perfect) paradigmatic proximations of the empirical universe (McGettigan, 1999, 2002). Therefore, humans have at their disposal the necessary cognitive mechanism, i.e., moments of truth, through which to do good science by taking gradual but confident steps toward a broader understanding of the infinite Truths that govern the universe.

Darwin's Spectacular New Truth

Thus, in contrast with Thomas Kuhn, I would answer the question, "Did Darwin tell the truth?" in the affirmative. I believe that Darwin told the truth because, in order to develop his theory of evolution, Darwin had to redefine reality. Once again, Darwin began his career as an uncritical creationist. After embarking on the HMS Beagle, however, Darwin encountered anomalies that severely tested his faith, e.g., fossils in the Andes, "mocking thrushes" and giant tortoises in the Galapagos, etc. So long as he remained under the sway of creationism, Darwin banished such subversive facts to the margins of his imagination. But still his doubts lingered, and with each new anomaly that he encountered, Darwin's creationist convictions gradually gave way to his subterranean evolutionary conjectures. This continued until Darwin experienced a moment of truth. Only Darwin could say precisely when his eureka moment occurred. It could have happened when John Gould enlightened Darwin about the taxonomy of Galapagos finches, or it might have occurred after reading Malthus' essay on population, or perhaps Darwin experienced his moment of truth during a period of quiet reflection in his study. Regardless, arrive it did.

Following his moment of truth, Darwin constructed a new, more verisimilar (Popper, 1965, p. 393) perspective than the creationist paradigm: evolution explains a broader range of facts in a single logically-coherent framework than the creationist paradigm. As such, I argue that Darwin's evolutionary paradigm is more truthful than creationism because, not only did Darwin redefine reality, but his theory corresponds with the facts to a greater degree than creationism.

Having said that, I must emphasize that, although it is an improvement on creationism, Darwin's evolutionary paradigm does not offer a perfect representation of the absolute Truth. Indeed, a crucial imperfection in Darwin's theory lay in his inability to understand how individuals pass advantageous mutations on to their offspring. It took Gregor Mendel's (Mawer, 2006) work with genetics and later discoveries revealing the existence of DNA (Watson, 1968) to shed light on that particular mystery. Thus, no paradigm is perfect, they can all be improved. However, through the process of redefining reality, Darwin was able to develop a more truthful framework through which to understand the origin and evolution of life. As such, Darwin's achievement is often rightfully acclaimed to be one of the greatest feats in scientific history.

Pleased as Darwin must have been with his scientific achievements, he must also have been disappointed by their public reception. Darwin was not a lover of controversy. Thus, he may have dreamed of introducing evolution to the world without stirring up a huge wave of public rancor. But it was not to be. Darwin's ideas were so revolutionary that they were bound to provoke widespread intellectual crisis. Indeed, on a somewhat more modest scale, the same is true for all new truths. Redefining reality is predicated on a process that attacks the status quo and introduces radically mind-altering alternatives. As a result, good scientists, like Darwin, Galileo, Einstein, are all of necessity a bunch of rebels.

In the next chapter, I will further refine the definition of good science while also explaining why good scientists must be rebels. The process of producing truth requires that agents counteract ideological biases that are part and parcel of the status quo. In brief, this is how science progresses. Further, this process also describes a means by which agents can introduce social change while also laying the foundation for new and enduring social realities.

Chapter 3
Good Science for Rebels

In the 1950s, a new era dawned in the United States. The United States emerged from World War II as the wealthiest, most powerful nation in the world. Prior to WWII, the United States had languished deep in the throes of an intractable economic Depression. Following WWII, the United States not only had sufficient funds to rebuild Europe (Hogan, 1987), but the United States also financed an unprecedented expansion in its own middle class (Lipset and Bendix, 1992). Though the emergence of the USSR as a nuclear antagonist darkened the spirits of many Americans, there was still plenty to celebrate. As the United States grew into its new role as the leader of the free world, rosy assessments of America's ascendancy proliferated (Galbraith, 1958; Parsons, 1960; Rostow, 1960). All the indicators suggested that the United States would be the next great empire. As it embarked on its thousand year reign, accolades flowed like champagne at the christening of an ocean liner. Democracy had triumphed.

Widespread as celebrations of American success might have been, nevertheless, there were also some skeptics. One of those dissenters was a sociologist named C. Wright Mills. During a time when questioning the virtue of the United States could easily invoke McCarthyist censure (Fried, 1997), Mills had the audacity to propose that America's newfound supremacy posed perhaps the gravest threat to democracy ever. Though it ruffled feathers to say so, Mills argued that, by the 1950s, United States democracy had become little more than an illusion (Mills, 1956).

The Military-Industrial Complex

The United States was a late entrant into WWII. There were a number of reasons for this delay. Wars are extraordinarily expensive undertakings and have more than once decimated the treasuries of even the wealthiest nations. In the late 1930s, the United States was still mired in the Great Depression and there were fears that full engagement in the war might deliver a death blow to the ailing American economy. However, as direct involvement in the war became inevitable, leaders in the federal government, military and major corporations inaugurated a strategic partnership that revolutionized the political-economics of war-making. This partnership has since come to be known as the military-industrial complex (Turse, 2008).

The goal of the military-industrial complex was to unite the strengths of three pivotal organizational groups: the military would design a winning war strategy, the federal government would provide coordination and finance, and corporations would manufacture the necessary war machines for, and this was the clincher, *a guaranteed profit.* How could any self-respecting corporation refuse? Not only was the federal government willing to underwrite prohibitively expensive R&D funds, but the feds also promised to be the best customer that profit-hungry companies could ever desire: purchasing vast quantities of commodities to furnish the bulk of the Allied war effort. Indeed, the military-industrial complex created enormous advantages for each of the organizational entities involved: corporations raked in their profits, the military got their cutting edge war machines, and the federal government benefited from all of the augmented military, economic and political clout that it could wield on the world stage. Indeed, the military-industrial complex generated such an efficacious organizational synergy that it transformed the United States into the single most powerful juggernaut the world has ever seen.

Though it was impossible to dispute the achievements of the military-industrial complex, Mills pointed out that the convergence of three such powerful organizational groups had the result of creating an entirely new, anti-democratic power structure in the United States (Mills, 1956). According to Mills, in the post-World War II era, the United States was no longer governed by democratically elected representatives, rather Mills argued that the nation had fallen under the sway of "power elites." Power elites are a shadowy group of individuals who have high-level connections to the three most powerful organizational clusters in the United States (i.e., the federal government, military, and Fortune 500 corporations) and who wield power largely without public oversight.

Critics were quick to assail Mills. While some attacked Mills' politics, others preferred to challenge what they perceived as the weaknesses in Mills' scientific philosophy. One such person was Robert Dahl (1961), who argued that Mills could only have arrived at his misguided conclusions by failing to embrace the strictures of good science. According to Dahl, good scientists are empiricists. From Dahl's perspective, Mills' power elite theory was a non-empirical and, thus, anti-scientific theory because it was based on a purely conceptual definition of power. That is, Mills argued that power elites corrupted democracy by wielding invisible influence over the political system. Rather than seeking the limelight and exposing their political machinations to public scrutiny, Mills argued that power elites actually amplified their influence by intentionally avoiding observable exercises of power. Employing their organizational clout, Mills argued that power elites exert informal, behind-the-scenes influence over policy-makers and thereby frame the context in which political debates take place. As such, power elites can manipulate the political landscape to their unique advantage without appearing to be directly involved in the political process, e.g., orchestrating increased military funding during the relatively peaceful post-war era (Eisenhower, 1961). Consequently, Mills concluded that power elites exercise power most efficaciously by sustaining the illusion of the democratic process, while simultaneously subverting that system to serve their ulterior objectives.

In an effort to challenge Mills' anti-empirical definition of power, as well as Mills' jaundiced view of the United States political system, Robert Dahl conducted a study of the democratic representational process in New Haven, Connecticut (Dahl, 1961). While Dahl was prepared to admit that representational politics were often cunningly nuanced, nevertheless, he insisted that good scientists could observe power empirically by scrutinizing the processes through which politicians make key decisions. In certain respects, Dahl's insistence on treating power as an empirically-observable phenomenon was reminiscent of Popper's assertion that truth corresponds with the facts. From Dahl's perspective, if one wished to learn the truth about power, one needed to define power in such a way that it could be observed and analyzed as an empirical fact.

Thus, Dahl defined his "intuitive idea of power" as follows: "*A* has power over *B* to the extent that he can get *B* to do something that *B* would not otherwise do" (1957, p. 80). Although this definition might sound a bit abstract, Dahl argued that antagonists exercised power in observable contests, e.g., congressional representatives who disagree over healthcare reform, war funding, immigration policy, etc. Democratic political debates, from Dahl's perspective, should be viewed as exercises of political pluralism. That is, in a democracy, political debates tend to be noisy and contentious because they usually involve a large number of participants who represent a wide spectrum of opinions. Yet, Dahl argued that in spite of all the pluralistic chaos, it remained possible to identify the power brokers who wielded the greatest influence. Dahl argued that the most powerful politicians were not necessarily the noisiest debaters. Rather, Dahl believed that the real power-brokers were those who won hotly contested decisions on the most crucial issues. Thus, in Dahl's study, the struggle to win key decisions represented observable, factual exercises of power. In essence, from Dahl's perspective, political power is roughly equivalent to a tally of key political victories. Thus, Dahl dismissed Mills' dubiously scientific speculations about the power elite because Mills incorporated explicitly unobservable criteria (i.e., shadowy political maneuvering by elites) into his analysis of the democratic political system.

Illusions of Power

For his part, Mills argued that Dahl's focus on observable representations of power permitted only a limited analysis of politically-motivated decision making. Certainly, relying upon observation as a means to evaluate knowledge claims has a strong intuitive appeal. Human judgment is profoundly influenced by sensory observations: we invest faith in those things that we can see, smell, hear, taste, or touch, whereas phenomena that defy observation (e.g., the Abominable Snowman) tend to tax credulity. Nevertheless, power elite theorists (Mills, 1956; Domhoff, 2006) have long maintained that observable social phenomena are not reliable measures of truth. Indeed, power elite theorists assert that observable social phenomena are often deliberately distorted by powerbrokers for the explicit purpose of deceiving observers, e.g., Enron executives projected an illusion of prosperity that, until 2001, most onlookers accepted as truthful (McLean and Elkind, 2004). Therefore, according to this perspective, observable facts are often nothing more than cunning fabrications designed by the powerful to deflect attention from their nefarious undertakings. As a result, power elite theorists assert that those who maintain faith in the observable world cannot avoid being duped by the powerful, e.g., Enron made a killing as long as investors remained sold on the company's dissimulations.

In essence, Mills argues that power elites can often wield enormous organizational power for the express purpose of shaping the way that those who are subject to their power perceive facts, e.g., though one would expect cessations of conflict to net peace dividends, United States military-industrial spending often increases as much during peacetime as during wartime. Such willful manipulations of the facts are strikingly reminiscent of the difficulties that Galileo encountered following the publication of his revolutionary astronomical observations. Although the facts clearly supported Galileo's heliocentric view of the cosmos, following an appraisal of the evidence that he had reported, the Inquisition insisted that Galileo reinterpret the facts to comply with the cosmology that had been endorsed by the Church. Thus, Mills argued that, due to the machinations of power elites, facts are often more likely to cloud than reveal truths.

To get beyond elite-generated distortions, Mills argued that observers needed to employ a special form of insight, something that he referred to as a sociological imagination (Mills, 1959). According to Mills, a sociological imagination is a conceptual framework through which observers can ascertain the impact of invisible social forces—including the handiwork of power elites—on the landscape of empirical reality. Without a sociological imagination, Mills assured, scientists cannot avoid being misled by power elites. Therefore, to do good science, Mills asserts that it is the responsibility of sociologically-informed observers to do more than invest faith in the world of appearances. To find the truth, one must dig deeper. And to do so, one must employ one's abilities as a thinking, acting agent to counteract the cognitive limitations that tend to inhibit a critical and, thus, more truthful, analysis of empirical reality.

A Taste For What's on the Menu

Further complicating matters, radical power theorists (Lukes, 2005) have asserted that, while elite power-brokers may distort observable reality, an even more insidious form of cultural power tends to subvert the cognition of observers. This is accomplished, Lukes (2005) argues, by creating a dislocation between an individual's real and subjective interests—and induces what Marx describes as a false consciousness (Marx and Raines, 2004, p. 114). Essentially, this perspective proposes that a subtle but extremely persuasive form of power cultivates tastes (Bourdieu, 1984) in the minds of individual social actors. Such tastes tend to predispose individuals to pursue objectives that seem to be born of individual desire, but that are, in fact, inculcated by prevailing socio-cultural influences.

For example, the early twenty-first century cultural context of the United States tends to inscribe its citizens with tastes (Bourdieu, 1984; McGettigan, 1999, 2006; Ritzer, 2008) for private homes, automobiles, computers, credit cards, cell phones, and fast food. Generally, we do not view our appetite for such cultural products as the work of social coercion. However, if we were to be situated in a markedly different context, let's say sixteenth century Hudson Bay Inuit culture, then our desires would incline toward warm fur-lined clothing, well-constructed igloos, skin-covered kayaks, and raw seafood. In the context of pre-industrial Inuit culture, it would be preposterous to lust after Big Macs because the extant cultural system would neither exert the impetus to seek, nor include any of the essential means to fabricate such delicacies. Thus, the third face of power works as a remarkably effective macro-level social glue because of the way that it encourages individuals to apply themselves insatiably to the pursuit of those things that extant cultural systems are designed to provide. Conveniently, these selfsame forces facilitate the reproduction of the cultural landscape within which individuals are embedded (Willis 1977, Burawoy, 1979): our hunger for automobiles effectively sustains the infrastructure for an expanding automobile culture (Schlosser, 2001).

Thus, radical power imposes deductive biases in a manner that has certain similarities to the cognition-shaping influences of paradigms. Kuhn (1962) argued that paradigms tend to dispose their adherents to evaluate facts in a way that effectively perpetuates their orienting paradigm, i.e., paradigms persist by instilling enduring commitments to a dominating ideological framework through which individuals perceive, experience, and explain empirical reality. As such, viewing facts through the lens of a particular paradigm tends to reveal "truths" that amplify rather than erode conviction to established beliefs. For example, the endurance of the creationist paradigm has been due, in part, to the passion with which its stalwarts have defended its assertions. Indeed, the virulent attacks that creationist scholars launched upon early naturalists had a significant impact on Darwin. In addition to being rebuked by Richard Owen for suggesting that species might be mutable, Darwin was chastened by the severity with which Robert

Grant, a professor at Edinburgh University, was ridiculed for publicly espousing a Lamarckian evolutionary perspective (Grant and Estes, 2009, p. 228). Thus, long before he published *On the Origin of Species*, Darwin realized that challenging prevailing beliefs was among the most dangerous feats that scientists could attempt.

For some, being introduced to entirely new ways of perceiving the universe is a mind-boggling, but exhilarating experience (Toffler, 1970). However, others would rather sustain their faith in established belief systems than yield to the cognitive culture shock of a paradigm shift. Kuhn (1962) argued that faithful proponents of flawed paradigms routinely ignore facts that pose a conceptual threat to their preferred perspective, regardless of the eye-opening truths that such anomalies can teach. Indeed, paradigmatic devotion is often so strong that beliefs in an established perspective—even a paradigm that is undergoing a fatal crisis—can trump all other logical considerations. As Kuhn argued, the final death knell for outmoded paradigms often does not sound until a critical mass of supporters have finally gone to their graves.

Consequently, as a result of the pervasive cognition-distorting influences of radical power, many social theorists have argued that individuals are utterly incapable of observing truth. That is, if every cognitive reference point has been manipulated by cultural power, then any truths that an observer might identify must be either partly or wholly the product of manipulative social power. Michel Foucault explains:

> truth isn't outside power, or lacking in power: contrary to a myth whose history and functions would repay further study, truth isn't the reward of free spirits, the child of protracted solitude, nor the privilege of those who have succeeded in liberating themselves. Truth is a thing of this world: it is produced only by virtue of multiple forms of constraint. And it induces regular effects of power (1980, p. 131).

Thus, Foucault maintains that in every circumstance truth is an instrument of power. From this perspective, truth is a mechanism that is employed to achieve the positive goals of a political regime: truth encourages those that it influences to "do the right thing", i.e., conform to the will of established authority.

Good Science, Truth and Agency

Though Dahl criticized Mills for failing to ground his theory of power on empirically-accessible facts, Mills responded that facts and their associated interpretations are themselves instruments of power. In turn, radical power theorists have also illustrated that the manner in which observers interpret facts is never neutral—cognition tends to be shaped in such a way as to elicit interpretations that support dominant modes of thought. Thus, good scientists cannot interpret facts truthfully without a grasp of the invisible and multi-dimensional social forces that construct the landscape of what and how we observe.

Consequently, the challenge for good scientists is to find some way to observe facts truthfully. However, as the foregoing discussion emphasizes, to interpret facts truthfully, one must not merely observe facts; one must somehow find a way to consider facts independently of the perception-distorting influences of invisible power. Another way of saying this is that scientists need to be as objective as possible in order to produce truthful interpretations of the facts. Fair enough, but reasonable as that suggestion may seem, liberating one's thinking processes from the subtle influences of radical power is much more difficult than one might suppose.

In essence, radical power instills a set of tastes that encourage individuals to embrace prevailing cultural values, norms, interests and aspirations. Having been imprinted by our culture on such an elemental level—our personal passions (e.g., favorite foods, clothing, hobbies, career choice, occupation, etc.) are largely a product of cultural influences—it is difficult to determine precisely where and when individual-level thinking is distinguishable from socially-imposed prerogatives. Or, in the language of another sociological debate (Bourdieu, 1984; Giddens, 1984; Ritzer, 2000, pp. 521-552), social structures extensively *determine* the entire scope of human thought and action. This argument maintains that individuals are nothing more than the helpless, unwitting pawns of social power structures.

As such, social determinists (Althusser, 1969) have asserted that humans lack a definable quality of agency. In brief, agency can be understood as a form of individual-level intellectual creativity that operates independently of—or, indeed, contrarily to—the influences of social structural determination. Thus, agency is essentially a quality of mind that enables individuals to conceive original ideas and then act upon those inspirations. However, determinists have argued that, given the extent to which individual consciousness is shaped by socio-cultural influences, it is nigh on impossible to identify a type of individual-level thinking that operates independently of the influences of social power. In other words, there seems to be no way to distinguish individual-level initiative from the tastes that radical power has imprinted upon the minds of individual actors.

Agency as Opposition

Radical power arguably has an impact on every aspect of the self-building process. For example, Freud (1961), Mead (1934), and Habermas (1981) argue that the mind cannot develop in the absence of extensive social structuring processes. Nevertheless, I argue that it remains possible to identify types of thinking that emerge only as a consequence of contradicting social coercion.

For example, during the 1950s, the dominant discourse in the United States was celebratory. Citizens who were skeptical of United States ascendancy ran the risk of undergoing stigmatizing public inquiry (Fried, 1997)—which, in certain respects, was eerily reminiscent of Galileo's run-in with the Inquisition. Yet, in spite of those risks, Mills articulated a sharply critical analysis of United States democracy. In developing such a creatively rebellious perspective, Mills demonstrated his capacity for agency.

Drawing upon this example, I propose that agency can be understood as a form of inventive thinking that exhibits three distinguishing characteristics:

1. Originality—instead of rehashing conventional thinking, agents concoct novel ideas. For example, rather than embracing the dominant discourse, Mills' power elite theory offered a fresh, new perspective on the post-war political landscape in the United States.
2. Opposition—given that invisible forms of power tend to inspire conformity, acts of agency are predicated upon a definable element of individual-level opposition to societal influences. For example, Mills, complete with leather jacket and motorcycle, was not only a rebellious thinker, but he also projected the archetypical image of the 1950s rebel. In addition, although early in his career Darwin was a conventional thinker, Darwin ultimately broke out of his conventional mold and hatched a theory that audaciously challenged every major presupposition of the creationist paradigm.
3. Redefinition—in contrast with the assertions of social determinists, I argue that individuals are capable of challenging and eliminating the influences of invisible social power through a process of redefining reality.

The process of redefining reality—which, I argue, represents the clearest and most emphatic expression of human agency—is contingent upon intentional efforts to challenge the cognition-distorting influences of social power. Importantly, although agency is based upon ingenious individual-level opposition to societal-level cognitive controls, the consequences of agency are not entirely negative. Indeed, far from being a counterproductive exercise, expressions of agency represent a process through which individuals are able to actively generate moments of truth, while also playing a pivotal role in the production of beneficial social change.

Rebels with a Cause

Of necessity, the process of redefining reality is predicated upon an act of intellectual rebellion. Once again, redefining reality is a multi-stage process through which individuals—whether scientists or not—can challenge and overcome the consciousness-distorting influences of invisible power. Redefining reality generally begins when individuals notice a disjuncture between observable facts and established modes of explanation, e.g., a democracy that is supposed to serve the people, but that instead caters to the whims of the powerful. Due to their devotion to established modes of thought, some observers might ignore anomalies, or contrive a convenient explanation that sustains their belief in what is already known, e.g., United States democracy may be imperfect, but it distributes power pluralistically through its convoluted representational system. Alternately, more independent thinkers might treat such a dilemma as an opportunity to transcend the socially imposed barriers that constrain their understanding of observable reality.

The process of transcending socially imposed cognitive barriers often begins with a creative observation (e.g., "Hey! Why don't politicians ever follow through on their campaign promises?"). In some cases, individuals who are determined to make sense of the anomaly might follow up their observations by developing an individual-level intellectual challenge to established modes of understanding (i.e., it appears as though the United States democratic system is primarily designed to serve the interests of power-brokers). Such acts of intellectual rebellion tend to further erode the foundations of conventional thinking (i.e., "Based upon what I have observed, I no longer believe democracy in the United States serves the will of the people."). Finally, the culmination of the redefinition of reality process involves constructing an entirely new explanation that simultaneously explodes existing ideological boundaries while also advancing a more adequate description of the phenomena in question, i.e., the United States political system masquerades as a democracy, while functioning like an elite-centered oligarchy.

Thus, as the foregoing example illustrates, individuals occasionally demonstrate the requisite mental apparatus to make note of anomalies, develop creative new explanations for mysterious phenomena, and then overcome manifestations of social power that delimit their thought and action. Therefore, the thoughts and behaviors of individual social actors are not entirely determined by the invisible influences of social coercion. Instead, sometimes agents can creatively counteract the distorting influences of social coercion and, in so doing, generate moments of truth.

Once again, a moment of truth is an experience wherein individuals, via the process of redefining reality, are transported from an inadequate version of reality to a more satisfactory paradigm. These experiences may be considered relatively truthful in that they are generated through a process whereby agents systematically counteract the influences of invisible social power over their definitions of reality. Thus, Mills (1956) argues that people who confine their analysis of the United States political system to the realm of the observable (i.e., the words and deeds of elected politicians), cannot help but fall prey to artfully calculated illusions. From Mills' perspective, the observable activities of political actors in the United States are designed to provide a convincing impression that politics-as-usual lives up to the ideals of democracy. Yet, Mills argues that appearances are deceiving. While political representatives go through the motions of faithfully serving their constituents, shadowy operators work behind the scenes to ensure that politics-as-usual serves the interests not of the majority, but of a privileged minority of power elites. Consequently, the truth is not defined by facts alone, rather the truth can only emerge as a result of a deeper investigation into the manner in which perception is often cunningly distorted by the interventions of social power. Therefore, it is in the process of counteracting the distorting influences of social power that it becomes possible for agents to experience moments of truth. However, it is important to emphasize that, even though agents may experience moments of truth, it would not be accurate to say that humans ever have or ever will formulate ultimate, universal Truths.

Once more, in agreement with radical power theorists (Foucault, 1977; Lukes, 2005), I argue that all knowledge systems exert some form of ideological power over the construction of knowledge. As such, this point returns to one of the key disputes between Thomas Kuhn and Karl Popper. Whereas Kuhn argued that the truth might exist, he believed it was not something that could ever be captured by science, nor could it guide the development of scientific theories: truth, being universal whereas paradigms are flawed and limited, implies that truth and human knowledge will remain forever disassociated. Popper, however, viewed the situation differently. Popper was unequivocal about the existence of truth; he was convinced that truth was a form of objective knowledge that existed independently of limited human perceptions and scientific theories. In essence, Popper's realist perspective asserted that the empirical universe existed independently of the minds of observers and it was shaped by difficult to comprehend laws of nature. Thus, Popper argued that there was an empirical universe 'out there' and, as such, it was possible for knowledge-seekers to reveal truths about that empirical universe. Of course, given the infinite scope of the universe, Popper believed that it was unrealistic to hope that humans might ever develop conjectures that would entirely compass universal truths. Similar to Kuhn, Popper was convinced that human knowledge was flawed, and would forever remain so. However, Popper contended that the knowledge acquisition process was guided and informed by truth. Thus, Popper believed that human knowledge gradually progressed through the conjecture and refutation process toward more verisimilar, but eternally imperfect, representations of truth. Yet, although Popper was an emphatic believer in the role that truth played in the production of knowledge, due to his insistence that scientists should avoid verification at all costs, Popper did not specify the integral role that truth played in the production of knowledge.

As a remedy to Popper's underspecification of truth, I have argued that humans can generate increasingly truthful knowledge through a process of redefining reality. I agree with Kuhn and Popper that there is a dislocation between ultimate, universal truths and the knowledge constructs (i.e., paradigms, conjectures, theories, etc.) that humans create. Recognizing the imperfections that are intrinsic to human knowledge constructs, I argue that via a creative intellectual process (i.e., redefining reality) individuals have the capacity to remedy imperfections in the structure of established knowledge in a manner that is both informed by, and productive of truth.

Thus, given the role that truth plays in the knowledge production process, I argue that it is essential for good scientists to be rebels. First of all, good scientists must be rebels because, as Mills argues, observers are more likely to be deceived than enlightened if they accept empirical facts at face value. Rather than interpreting empirical facts as they appear, good scientists must interrogate facts not only for the truths that they might reveal but also for the deceptions that, when examined uncritically, they can perpetuate. As power theorists have pointed out, truth regimes have a tendency to supply explanations for empirical facts that have the consequence of generating additional support for established belief systems. Thus, good scientists must challenge the status quo by investigating beyond the world of appearances—as Galileo accomplished with his telescope and as Mills achieved with his sociological imagination—in order to develop new explanations for facts that challenge conventional beliefs. In turn, to make sense of the truths that lie beyond the realm of superficial appearances, good scientists must engage their capacity to act as agents. According to the definition that I have developed above, agents are not only creative thinkers, but they must also be defined in terms of their capacity to counteract, or rebel against, the influences of established power structures that limit or distort their ability to generate new explanations for anomalies.

Finally, good scientists are rebels in the truest sense of the term because, in their effort to seek truth, good scientists have often instigated social change. That is, by redefining reality, scientists have reframed the boundaries of understanding and introduced entirely new facts (e.g., new planets, new biological processes, new dimensions of socio-political analysis, new frontiers in space flight, etc.) into the known universe. Thus, redefining reality not only transforms the known universe in a conceptual sense (i.e., from geocentrism to the expanding universe), but truth-seeking has also brought about very literal modifications to lived reality, e.g., aviation has become second nature since the Wright Brothers' first flight, and, thanks to Darwin, scientists have mapped and even begun to tinker with the human genome. Thus, the search for truth has very literally transformed the boundaries of reality. Indeed, as long as good scientists persist in their search for truth, then those socially-situated, rebellious agents will continue to stimulate social change in dramatic and often unexpected ways.

Socially-Situated Agency

The version of "redefined truth" that I advocate asserts that no single person will ever arrive at an ultimate representation of Truth. Instead, humans can access narrow, momentary glimpses of truth through the process of transitioning from outmoded to improved definitions of reality. Once again, as scientists (and private citizens) it is essential to embrace coherent truth standards in order to establish a foundation upon which to attack inadequate ideas and replace with them with better, more truthful ideas. In denying the existence of truth standards, one foregoes any rational basis upon which to rebuke quacks, e.g., tyrants who proclaim that inferior people should be exterminated, barstool physicists who profess to know more about relativity than Einstein, or sociologists who contend that all forms of knowledge are equally valid.

Agents can pursue truth through the process of redefining reality; however, in every case it remains up to individual observers to evaluate the veracity of knowledge claims. For example, even the most widely accepted scientific paradigms are, and should be, subjected to intense criticism (Behe, 1996; National Academy of Sciences, 2008). An environment that invites criticism of even the most popular theories—regardless of whether or not we esteem dissenters' viewpoints—is crucial to the process of progressively and legitimately redefining reality. In other words, dissent is an acid test through which to interrogate good ideas and obliterate bad ones. Once again, no theory produced by humankind either has, or ever will capture the Entire Truth. Indeed, precisely because of that limitation, the notion of redefined truth is an essential means through which to emphasize that even relatively truthful ideas can and should be supplanted by better ideas.

Indeed, given the foregoing discussion about thc limitations of human-made knowledge systems, I propose that eternally provisional, but increasingly proximate paradigms emerge from a negotiation process between:

1. Ingenious, redefining human minds
2. Observable (and, in particular, anomalous) empirical phenomena
3. Established paradigms

Thus, the process of redefining reality implies that good scientists can only obtain a competent understanding of the empirical universe by acting as agents from within the context of coercive social reality (McGettigan, 1999, 2001, 2006). Truth-seekers must confront the invisible influences of social power in order to effectively grasp the complexities of the simultaneously contradictory and complimentary relationship between agents and the manipulative social environments that they inhabit—and, thereby, generate scientific paradigms of increasing veracity.

Intellectual Evolution vs. Extinction

Of necessity, truth must remain an evolving target. As the boundaries of the known universe expand, the paradigms that extend scientific frontiers must keep pace. Humbling though it may be, the surest route to scientific progress is to assume that the existing state of knowledge is lamentable. Whether we like it or not, the grandest truths that we cling to today will eventually seem as quaint as the once firmly held conviction that earth was flat. Indeed, the mere suggestion that any scientific discipline may be nearing an ultimate theory (Greene, 2003; Hawking, 2002) is exceedingly curious. No scientific discipline can brag of greater accomplishments than physics. Nevertheless, given that physicists currently acknowledge that the universe is permeated by vast quantities of inscrutable dark substances (Clegg, 2009; Overbye, 2006), I believe it is premature, to say the least, to assert that ultimate theories lie just beyond our ken. In fact, rather than being on the verge of an ultimate theory, I would assert that such fragmentary comprehension of the dark substances that comprise the bulk of our universe augers an impending paradigm shift of epic proportions. When such a paradigm shift inevitably occurs, those who have predicted that "the end is near" will be wearing egg on their faces. Whereas, everyone else will benefit from science having, yet again, taken one small, but tantalizing step toward compassing the universe's far-flung mysteries.

Without doubt, the greatest threat to science is the presumption that any particular paradigm might somehow encapsulate the final, ultimate Truth. In every case where humans have claimed ownership of final truths, agency has been forced to cower in the face of bloody-minded ideology (Barnett, 2006; Popper, 1950). If scientists are ever unwary enough to presume that their intellectual journey has arrived at its final destination, then in that very moment the aspirations of good scientists, the quest for truth and intellectual integrity itself will suffer a painful, premature and unnecessary extinction.

The life of science is dependent upon an enduring commitment to intellectual evolution. That being the case, scientists can only preserve their enterprise by maintaining a steadfast commitment to human agency. In other words, scientists must always maintain a stronger commitment to doubt than to any of the certainties touted by prevailing paradigms. Toward that end, scientists must stubbornly seek anomalies that will be certain to undermine the paradigmatic beliefs upon which cherished truths and distinguished careers have been founded. This is essential not purely for the purposes of tearing down scientific aspirations—as postmodernists have been wont to do. Rather, the goal of accumulating anomalies is to force confrontations between established paradigms and the truths that transcend those paradigms. Science is at its best when it remains singularly committed to the goal of evolving paradigms and, in so doing, focusing scientists' unwavering aspirations on the truths that extend perpetually beyond their wildest imaginations.

In the next chapter, I will introduce the concept of problematics. Problematics can be understood as exceedingly challenging goals that require knowledge-seekers to invent new facts and, ultimately, to redefine reality in order to achieve their far-fetched objectives. As such, problematics represent a means through which to proactively challenge the paradigmatic intellectual constraints that tend to stymie constructive intellectual progress.

To illustrate the invaluable role that problematics can play in generating intentional intellectual progress, I will examine the space race that unfolded in the 1960s. By advocating the goal of a pre-1970 moon landing, JFK posed the United States with a seemingly impossible problematic that ultimately enabled the United States to redefine itself as a nation.

Chapter 4
The Sky is Not the Limit

Sputnik changed everything. The fact that the USSR launched the first human-made satellite into orbit constituted a major blow to the American psyche. World War II helped make the United States the leader of the free world. With the detonation of atomic bombs in Hiroshima and Nagasaki, the United States brought an emphatic conclusion to the war (Rhodes, 1986). Dreadful as those nuclear attacks may have been, they signaled that, via its ultra-secret Manhattan Project, the United States had arrived as the world's pre-eminent military and technological power.

Yet, the United States' singular position as the world's only possessor of nuclear weapons did not last long. On August 29, 1949, the Soviets detonated RDS-1 (a.k.a., Joe 1), their first atomic bomb. In that instant, the United States became just one member of an expanding club of atomic powers. Worse yet, it was the Soviets, the United States' arch nemesis, that had been the first to match that most significant scientific achievement. And, most troubling of all, the Soviets had done so far sooner than anyone had imagined possible (Rhodes, 1995).

The Enemy of My Enemy

Although the United States and the Soviet Union had teamed up as allies during WWII, the relationship had never been a warm one. United States suspicions of the Soviets were rooted in fundamental philosophical differences. That is, the Soviet Union was founded on a Marxist political program that openly deplored many of America's most cherished principles. As a result, it was nothing short of a miracle that the United States and Soviet Union managed to cooperate during WWII. However, when WWII ended, the strategic partnership between the United States and Soviets rapidly dissolved. Indeed, when the Soviets tested Joe 1, they touched off an entirely new military struggle. The nuclear arms race that ensued between the United States and the Soviet Union gradually swept nations all over the globe into two opposed and competing camps (Gaddis, 2005).

Without doubt, the cold war was the most technologically-sophisticated war ever waged, and it was also the most irrational (Rhodes, 2007). Humans had finally managed to create a class of weapons that were destructive enough to threaten all of humanity, both friend and foe alike. Indeed, in the aftermath of the Hiroshima and Nagasaki bombings, a debate broke out in the United States about whether there was any need to continue developing nuclear weapons. Those opposed to augmenting the United States arsenal were joined by a most unexpected ally, Robert Oppenheimer (Bird and Sherwin, 2005). Although he had been the scientific leader of the Manhattan Project, following the conclusion of WWII, Oppenheimer also became a determined opponent of nuclear proliferation. Oppenheimer argued that he and many of his colleagues had participated in the Manhattan Project largely because the bomb had represented the greatest scientific challenge of their time. However, having managed to construct a working atomic bomb—as well as having witnessed its horrific destructive power—Oppenheimer decided to wield his personal and political influence to oppose post-war nuclear proliferation. Why, Oppenheimer wondered, should the United States build weapons that posed as grave a threat to the United States as they did to our enemies?

Indeed, partly due to Oppenheimer's opposition, in the years following WWII there was limited progress on nuclear technologies in the United States. Oppenheimer made the persuasive argument that the Manhattan Project had propelled United States nuclear capabilities so far ahead of its enemies that they would be unable to catch up for decades (Rhodes, 1995). The United States could thus sit tight on its technological advantages and reassign the vast resources supporting atomic research to other peacetime projects. However, the situation changed when the Soviets detonated Joe 1. Despite Oppenheimer's optimistic predictions, it turned out that the United States did not enjoy as great a technological advantage as previously imagined. To be fair, it soon came to light that the Soviets had closed the atomic weapons gap with the help of an expansive espionage network (Hornblum, 2010).

Nevertheless, the Soviet Union's unexpected detonation of Joe 1 immediately made nuclear weapons research a top priority in the United States. One consequence of this shift was that public trust in Robert Oppenheimer eroded (Bird and Sherwin, 2005). Indeed, Oppenheimer's opponents even mounted a rumor campaign designed to tarnish Oppenheimer's image (Teller and Schoolery, 2001, p. 372). Over the years to come, not only would Oppenheimer's scientific judgment be called into question, but so would his patriotism. In the end, due to a cloud of anti-communist suspicion, Oppenheimer would tumble from the heights of Manhattan Project leadership to having his security clearance revoked (Pais, 2006). In sum, a tragic epilogue to a life devoted to science and service to his country.

Having rededicated itself to nuclear dominance, the United States soon raised the bar in the global arms race. On November 1, 1952, the United States detonated Ivy Mike, the first nuclear fusion bomb (Van Dorn, 2008). Whereas the explosive power of WWII-era atomic bombs had been measured in kilotons, post-war fusion bombs detonated with megatons of explosive force. Staggering as that advancement may have been, less than one year later, the Soviets detonated their own fusion bomb (Norris and Arkin, 1993, p. 48). Thus, in spite of America's post-war progress in bomb-making, the technology gap between the United States and Soviets was shrinking rapidly.

As the 1950s wore on, there were additional signs that the Soviets were not only narrowing the technology gap, but they might actually be bounding ahead. Indeed, America's worst fears were realized when on October 4, 1957, the USSR launched Sputnik, the very first artificial satellite (Dickson, 2003). Although Sputnik was little more than a beach ball-sized radio transmitter, Sputnik represented a watershed moment in the US-Soviet race for global supremacy. For the first time, the Soviets had achieved a landmark technological goal before the United States (Cadbury, 2006). As every 96.2 minutes Sputnik whizzed overhead, all the while chirping its maddening birdsong, anxiety intensified among Americans. Though Sputnik posed no direct threat to America's national security, its mere existence was a nagging reminder of Soviet ascendancy. Not only had the USSR beaten the United States into space, but as the Soviet's dratted satellite streaked overhead, it became abundantly clear that the United States lacked the ability to respond tangibly. If the Soviets could put satellites in orbit, then what else might they be up to? Imaginations ran wild. Lacking a credible counterstrike capability, the United States suddenly found itself at the mercy of its cold war nemesis. Americans would have no peace so long as the Soviets enjoyed a technological advantage over the United States.

From Bad to Worse

As it turned out, Sputnik was only the first of a seemingly interminable series of Soviet precedents in outer space. On November 3, 1957, barely one month following the launch of Sputnik, the Soviets fired Sputnik 2 into orbit. Sputnik 2 was noteworthy for carrying a dog named Laika, the first living animal ever launched into space (Owen, 2004). More bothersome still, on April 12, 1961, the Soviets claimed the next great prize in the space race by launching the first human being, Yuri Gagarin, into space (Schefter, 1999).

Still, the Soviets would not lay sole claim to all of the major feats in space flight. A mere twenty-three days after Gagarin's groundbreaking flight, the United States followed suit by launching the first of its Mercury astronauts, Alan Shepard, into space. While Shepard's first flight was celebrated as a history-making achievement, nevertheless, one also had to acknowledge that, beginning with Sputnik, the United States always seemed to show up a day late and a dollar short in the race for technological supremacy.

Recognizing both the practical and political dangers of lagging behind the Soviets, President John F. Kennedy decided to announce a bold new goal for the United States space program. On May 25, 1961, JFK stood before a joint session of Congress and made an extraordinary commitment (French and Burgess, 2007). In 1961, with Sputnik whizzing by overhead, dominoes tumbling in the developing world, and swords rattling in Cuba, the United States appeared to be woefully outflanked by its adversaries. Yet, rather than buckling under the weight of those combined difficulties, JFK stunned the world by announcing that, instead of donning the label of international has-been, the United States would reassert its claim to international dominance by pursuing a goal that lay beyond any other nation's wildest dreams. The United States would outdistance its rivals by aiming higher: the United States would shoot for the moon. Thus, Kennedy made the jaw-dropping announcement that, before the decade was out, the United States would send a man to the moon and return him safely to earth (Orloff and Harland, 2006).

Perhaps the most striking aspect of JFK's announcement was that, as of May 25, 1961, the prospect of landing humans on the moon was pure science fiction. Quite literally, as of 1961, the technology did not exist to achieve Kennedy's far-fetched goal. Indeed, it had taken a monumental effort just to launch Alan Shepherd a few miles above the earth's surface. In contrast with Yuri Gagarin, Shepard had not even orbited the planet. If the truth be told, there was even some doubt about whether Shepard's voyage had actually constituted a space flight; the Soviets insisted that Shepard's suborbital rocket ride had not even boosted him above the uppermost reaches of the earth's atmosphere (Chaikin, 2007). Thus, the United States scientific community's knee jerk reaction to Kennedy's newly revealed goal was, "It can't be done." In 1961, the United States simply did not have the technological know-how to send astronauts deep into space.

No one knew better than the scientific community how ill-equipped the United States was to launch a manned moon flight initiative in 1961. Indeed, had NASA scientists attempted to use early 1960s technologies to shoot for the moon, the missions would have failed (Erickson, 2005). Still, impossible as it may have seemed, Kennedy had thrown down the gauntlet. As a visionary leader, Kennedy had decided that, no matter what obstacles might lie in the path, the best chance of building a brighter future for the United States was to lift the nation's eyes, hearts, and minds toward the moon. The choice was ours. Americans could wallow in the muck and mire of self-doubt, or they could take flight with the dreams of a courageous leader. Thus, for NASA to shoot for the moon, they would have to find some way to achieve the impossible. In the end, they were able to do precisely that because, as humans had done so many times before, NASA scientists consciously revised the boundaries of the possible, i.e., they invented new technologies, created new organizations, developed new spaceflight procedures and, in the end, they re-invented reality.

Though the history of lunar space flight is rife with disasters and fateful near misses, nevertheless, on July 20, 1969, America realized Kennedy's vision

(Harland, 2007). The United States became the first—and still remains the only—nation to send its citizens to the moon. Yet, apart from that important distinction, what was the real significance of this landmark achievement?

Super-Adaptable Apes

The fate of many creatures tends to be sealed, or determined, by the circumstances that they encounter in their environment. The dinosaurs dominated the planet for millions of years until a catastrophic environmental shift brought about their sudden demise (Horner and Gorman, 2009).

On numerous occasions, Karl Popper (1999) made the point that "all life is problem solving." By that, Popper meant that survival for every life form is contingent upon efficacious problem solving: where to find food, how to stay warm, how to secure a mate, how to avoid predators, etc. Generally speaking, most creatures rely on their genetics to solve problems, e.g., furry little mammals were better suited than dinosaurs to capitalize on shifting environmental conditions at the K-T boundary (Koeberl and MacLeod, 2002, p. 261). In other words, for most creatures, problem solving is a purely biological process. Either creatures are equipped with the necessary genetic attributes to survive in a given ecological context, or they are not. Thus, lacking the necessary genetic attributes to withstand the K-T environmental transition, the dinosaurs went extinct.

However, for humans, problem-solving is not solely determined by the match between genetics and environmental conditions. To a much greater extent than any other creature, humans solve problems cognitively. Certainly, human cognition is a genetically evolved attribute. If humans have big brains, it's only due to long-term evolutionary processes that have conferred advantages on brainier hominids (Lynch and Granger, 2008). Thus, genes matter and the evolutionary process applies to all living things. That said, cognition has permitted humans to circumvent the constraints of purely biological adaptation. In other words, humans have become cognitively "super-adaptable." Super-adaptability can be understood as the unique human capacity to develop intentional, non-random solutions to environmental problems at a pace that is far more rapid than the biological evolutionary process.

Just like every living creature, in their effort to survive humans routinely encounter environmental constraints (i.e., too little food, water, space, warmth, etc.). However, humans have been able to overcome such obstacles by super-adapting. In a very literal sense, this means that, upon encountering empirical or conceptual obstacles that hamper the achievement of a particular goal, humans have employed their capacity to imagine alternative realities in order to achieve their goals, e.g., conquest of the moon. In the process of super-adapting, humans routinely invent new facts and modify the parameters of reality. Where once, there were simply too many environmental barriers to seriously contemplate space flight, over the past half-century super-adaptable humans have gradually made extra-terrestrial travel commonplace (Carroll, 2009).

As explained in chapter three, redefining reality is a process through which humans develop new, more truthful conceptual schemes for explaining facts. Creatures that lack the capacity for redefining reality can only mount biological responses to environmental challenges, i.e., they are either biologically fit for an environment or they aren't. Humans, on the other hand, have the ability to approach environmental challenges as intellectual problems (Diamond, 1998). Thus, the disconnect between environmental conditions and innate biological attributes (e.g., bare skin in a cold climate), represents a problem that stimulates super-adaptable apes to redefine the status quo. In this instance, humans overcame the limitations of equatorial biology by developing cognitive solutions (i.e., utilizing fire, fashioning garments out of furry animals, living in shelters, etc.) to the problem of bare skin in colder climates. Consequently, instead of perishing as they made forays into non-native climes, humans developed super-adaptable, intellectual solutions to the many environmental challenges that they encountered (Gould, 1997, p. 220). In doing so, humans demonstrated their agency, or a unique ability to invent novel solutions to environmental constraints.

As a result of their super-adaptability, humans have been able to assert a much greater degree of control over the environment. In other words, humans have employed their intellectual abilities to construct increasingly synthesized living experiences. For example, instead of being limited by the food resources that are readily accessible in the environment, humans have developed various forms of agriculture as a means of augmenting the quantity and variety of food resources (Pollan, 2006). Not only do farmers produce bumper crops of food, but they cultivate agricultural products that have been optimized to suit consumer tastes (Schlosser, 2001).

Indeed, humans have become so adept at modifying the environment that in post-industrial societies the bulk of day-to-day experience tends to take place within elaborately synthesized environments. For example, rather than living in direct contact with nature, humans have developed increasingly elaborate synthetic envelopes, such as:

- Dwellings with built-in climate control, running water, artificial light, and processed foods—all of which are designed to increase comfort and, thus, decrease immediate survival challenges
- Synthesized entertainment (TV, radio, cinema, streaming video), communication (phones, verbal and text messaging), information production and exchange (computers, Internet), and virtual interaction (social networking, tweeting) enable humans to become more deeply embedded in "unnatural," but highly advantageous cyber realities

Consequently, many humans have begun to lead lives that are almost entirely disconnected from the natural environment. Indeed, Jean Baudrillard (1989, 1994) argued that Americans have become so thoroughly cocooned in synthesized realities that it can be difficult to distinguish Americans from the simulacra in which they are embedded. Admittedly, Baudrillard's observations often border on the absurd; however, he does make the valid observation that Americans are becoming increasingly wedded to information technologies—so much so that many find it distressing to unplug from their digitally-mediated realities (Dretzin, 2010). Certainly, there are reasons to be concerned about the negative health effects that can result from leading overly synthesized lifestyles (Louv, 2008). Nevertheless, in contrast with Baudrillard's fatalistic perspective (Gane, 1991), I argue that there is much to be gained by constructing increasingly synthesized realities.

Not-So Impossible Dreams

Artfully constructed fantasies often have a significant impact on the course of human events. For instance, in the nineteenth century, Jules Verne concocted outlandish visions of the future that thrilled his many readers. It would be an overstatement to suggest that Verne's science fiction fantasies laid the groundwork for the historical events that were to follow. Still, I believe it is fair to say that some of Verne's diehard readers have made a concerted effort to transform his fantasies into reality. For example, in Verne's day, a fully electronic submarine was a work of pure imagination. However, in the century that has followed, engineering marvels that bear striking similarities to the fictional submarine that Captain Nemo commanded have confidently plumbed the depths of the seven seas. Indeed, it is noteworthy that the first nuclear submarine in the United States fleet was named the Nautilus (Seelhorst, 2003). Thus, in some respects, the future is often inspired by fantasies.

Of course, not every fantasy has a transformative impact on reality (Montandon, 2008). There are plenty of far-fetched fantasies that will never be more than works of pure imagination, e.g., super-heroes that can lift trains over their heads, fly unaided through the skies, or blast laser beams from their eyes. On the other hand, fantasies can sometimes inspire humans to transform their relationship with the environment and, in so doing, radically redefine the boundaries of the possible.

As I mentioned above, in 1961, JFK's plan to send astronauts to the moon, was a fantasy. In fact, his proposal was very similar to the simulacra that animate Hollywood movies: an imaginative vision of an alternate reality which, in this case, resembled the plot for Stanley Kubrick's film, *2001: A Space Odyssey*. In 1961, technologies to land astronauts on the moon existed only in imaginations (Thimmesh, 2006); and, some believed, only in hyperactive imaginations at that. Yet, fantastic as Kennedy's proposed lunar landing may have been, his wildly optimistic proposal created a set of challenges—or, what I will refer to as a problematic—that humans happen to be uniquely well-adapted to resolve. That is, ever since pre-humans began wielding sharp sticks and stones, humans and their ancestors have been purposefully redefining the course of events and the fabric of empirical reality.

Paradigms vs. Problematics

As explained earlier, paradigms are broad conceptual constructs that are intended to render the known universe intelligible. Paradigms play an important role in intellectual advancement by providing a framework within which to conduct normal scientific progress. When, at length, knowledge-seekers exhaust the limits of intellectual growth within a particular worldview, paradigms stimulate an altogether different level of intellectual innovation by inspiring the invention of revolutionary paradigmatic successors.

Problematics are somewhat different. Rather than providing a framework into which to incorporate facts, a problematic can be understood as call to action to resolve a particular scientific challenge, e.g. achieving heavier-than-air flight, or landing astronauts on the moon. Thus, problematics instigate an entirely different approach to truth-seeking than is typical of the paradigm revolution process. Paradigms endure so long as they provide adequate explanations for the majority of available empirical evidence. While problematics can often stimulate the process of paradigm revolution, problematics are neither defined nor limited by extant facts. Indeed, a problematic can be as far-fetched as its author's imagination can stretch, e.g., creating an electric submarine or even a time machine (Kaku, 2008; Nahin, 1999). Thus, a problematic poses an extraordinary and, thus, an imaginary goal toward which knowledge-seekers can aspire. Upon formulating the problematic, the "problem" for those who happen to be inspired by the problematic, becomes a question of how to transform the proposed fantasy into a reality.

Although scientists often insist that science progresses on the basis of facts (Popper, 1959), problematics play an integral role in the advancement of scientific knowledge. As mentioned above, instead of using facts as a departure point, problematic science identifies an appealing goal (e.g., heavier-than-air flight, moon landings, nanotechnology, artificial intelligence, etc.) and then sets about the process of creating the facts that are necessary to transform the problematic goal from a fantasy into a reality. In this context, truth is still generated through a process of redefining reality; however, the facts upon which new truths and new realities are founded are themselves inventions of human agency. Thus, truth must always be founded upon facts, however, via the problematic scientific fact-production process, there is almost no limit to the facts—and, by extension, the mind-blowing new realities—that creative human agents can invent.

Any Job Worth Doing

In 1961, JFK confronted a situation in which the facts seemed to suggest that the United States was falling hopelessly behind the Soviets. From Sputnik to Vostok, the Soviets had claimed every major technological achievement during the fledgling space race (Cadbury, 2006). Thus, for many, the evidence appeared to support the ominous conclusion that the United States was steadily falling behind the more enterprising Soviets.

As already discussed, any particular set of facts can be interpreted in a variety of ways. For instance, in the case of the space race, the facts seemed to suggest that the Soviets had achieved an impressive level of technological superiority over the United States. Yet, rather than embracing such a denigrating interpretation of the facts, President Kennedy decided to assert that the Soviets' apparent victories were naught but anomalies. From Kennedy's perspective, the truth could not be determined through a superficial examination of the facts. Rather, the truth about the United States could only be understood with reference to a modified reality that, Kennedy insisted, his nation would soon invent.

Whether Kennedy was an unflagging optimist, or whether he was engaged in a calculated exercise of political gamesmanship remains an open question. What is evident, however, is that Kennedy's bold vision for the United States space program provided Americans with an entirely new conceptual lens through which to appraise their potential. Although the logistical difficulties were monumental, Kennedy brushed aside the perceived difficulty of sending astronauts to the moon. In fact, Kennedy was of the opinion that, the more difficult the task, the greater its potential benefits:

> We choose to go to the moon in this decade and do the other things, not because they are easy, but because they are hard, because that goal will serve to organize and measure the best of our energies and skills, because that challenge is one that we are willing to accept, one we are unwilling to postpone, and one which we intend to win . . . (Address at Rice University on the Nation's Space Effort, President John F. Kennedy, September 12, 1962)

For Kennedy, the logistical challenges of sending astronauts to the moon were of secondary concern. Kennedy's primary goal was to create a new outlook for the nation. Whereas a climate of self-doubt had descended upon Americans, Kennedy strove to banish that mindset and replace it with an altogether different way of thinking; a new problematic if you will.

Problematic Innovation

The point of all this is to emphasize that humans have an extraordinary capacity to shape the future. JFK made the goal of a pre-1970 moon landing a top national priority. However, inspiring as JFK may have been as an orator, his verbal commitment alone was not sufficient to make his far-fetched dream a reality. Redefining reality is a process that begins with inventing new ways to reveal truths about the empirical universe. However, the act of imaginative reconceptualization is only the first step in the process of redefining reality. Once knowledge-seekers create a problematic vision, it is necessary to generate sufficient evidence to resolve the new problematic. Thus, Kennedy blazed a conceptual path to a new frontier, but it required the herculean efforts of an energized scientific community to make the dream a reality (Thimmesh, 2006).

Kennedy knew only too well how improbable his goal of a pre-1970 moon landing was. Had the United States been unable to transform JFK's dream into a reality—and there were plenty of points along the way where the hoped-for outcome was gravely in doubt (Erickson, 2005)—then history would have been uncharitable to Kennedy. No president wants to be perceived as a failure. Thus, Kennedy had to craft a new problematic for the United States that, while boosting the nation out of its doldrums, promised a real chance for success. Certainly, Kennedy was a gambler. In order to reach the moon, the United States would have to elevate space science to an entirely new threshold. However, Kennedy cleverly motivated the people of the United States in a way that energized not only their creative character, but, perhaps more importantly, their spirit of competitiveness. The American public threw their support behind the Apollo program in part because it was an inspiring technological goal, but, more importantly, because it created an opportunity to stick it to the Russkies. At long last, the United States was determined to beat the Soviets to a major goal in the space race. Even for Americans who lacked any interest in space exploration, that was a goal worth shooting for.

So, the United States turned its attention to the problematic of landing astronauts on the moon. Through an extraordinary commitment of time, resources and political will, the United States managed to progress in stages—from Mercury, to Gemini, to Apollo—toward its final goal. Step by step, NASA deployed solutions to the environmental obstacles that stood in the way of landing astronauts on the moon. In the end, NASA not only generated the facts that confirmed the truth of JFK's bold prediction, but NASA also redefined the relationship between humans and the cosmos. Where once humans were entirely earthbound, over the course of the 1960s, humans became extra-terrestrials.

In addition to equipping Americans with the ability to travel in the lifeless void of space, the struggle to win the space race also modified reality on the surface of the planet. Whereas the United States emerged from the 1950s at a technological disadvantage, the United States hurtled out of the 1960s as the unparalleled leader in space science. The Soviets could not come close to matching the United States' feat of landing astronauts on the moon (Chaikin, 2007). Ultimately, as the United States laid the groundwork to become the leader of the information society, the Soviets finally gave up the ghost. In 1991, George H. W. Bush declared that the cold war was over. The United States had won and, as a prize, Bush proclaimed America's ascendancy as the world's only superpower in a new world order. Arguably, the turning point in its long struggle against the Soviets was that defining moment when JFK dedicated the United States to victory in the space race.

The United States had succeeded in the space race and, ultimately, in the cold war by making intentional decisions about how to construct the future. This is not to say that people can create any future that they desire simply by putting their imaginations to work. As illustrated above, the future is modifiable to the extent that humans can invent viable problematics and then apply themselves to the arduous task of transforming their hoped-for objectives into reality. If dreamers are unable to tether their fantastic problematics to a firm foundation in empirical reality, then their dreams will remain naught but fantasies. However, for gifted visionaries, such as JFK, the future is a fantasy, but a fantasy that can be realized through the resolution of a particularly inspiring problematic.

Born to Adapt

Humans evolved from humble origins—from scavengers, no less (Rufus and Lawson, 2009). Even among scavengers, pre-humans would have numbered among the least intimidating creatures competing for scarce resources. Lacking long, sharp teeth, or deadly claws, pre-humans were poorly equipped to compete with their more well-endowed rivals. However, all of this changed when, due to some unknown inspiration, pre-humans began tinkering with technology.

The first tools were found objects (Reader, 1981). In other words, early tools were nothing more than the sticks and stones that pre-humans stumbled upon in their environment. The earliest tool-users would have lacked the ability to fashion tools, rather they would simply have utilized objects that were ready to hand in a moment of need; perhaps hurling a rock at a carnivore that was intent upon making a meal of the tool-wielder. In such a scenario, if the stone-thrower survived, the utility of the life-saving tool would likely have imparted a lasting impression. Such success could also have inspired tool use in other situations, e.g., driving ferocious scavengers from kill sites with a barrage of sticks and stones.

It is not perfectly clear under what circumstances pre-humans first began using tools (Dunsworth, 2007). However, once they did, it was a habit that stuck. Archaeological sites reveal that tools became more numerous, varied and complex as humans evolved. Indeed, there appears to be a striking correlation between the development of tools and the evolution of modern Homo sapiens (Diamond, 2006). As pre-humans crafted ever more cunning tools, the very physiology of the creatures that wrought them, and their brain cases in particular, changed dramatically. Modern Homo sapiens, the wisest of animals, evolved as a result of the unique advantages that tool use confers (Lenski, 1966; Lenski and Nolan, 2006).

Be Careful What You Wish For

I should emphasize that, under no circumstances am I suggesting that either genetics or environmental conditions are irrelevant in the struggle for human survival. Humans, like every other living creature, are products of the combined influences of their biology and environment. Genetic attributes, which certainly include the capacity for cognition, have made it possible for humans to capitalize on an extremely broad array of environmental opportunities. In addition, environmental factors have always played an important role in the disposition of human affairs: more than once, great civilizations have met their demise by failing to adapt to changing environmental circumstances (Diamond, 2005; Reisner, 1993). That said, rather than having their fate entirely determined by environmental conditions, the capacity to problem-solve cognitively enables humans to modify their environment in a way that other creatures simply cannot. For example, whereas many creatures solve problems associated with decreasing global temperatures by randomly evolving beneficial genetic attributes (e.g., more copious hair, downier feathers, thicker blubber, etc.), humans solve such challenges by problematizing the situation. What humans can't achieve with genetics, they orchestrate with reality-redefining problematics: flooding desert environments with life-giving water, building warm underground dwellings in Antarctica, constructing nuclear submarines that can sustain life under the seas, or assembling space stations in the extra-terrestrial vacuum, etc.

Of course, success usually begets new challenges, or as Popper argued, solving one problem generally evokes an entirely new set of problems (Popper, 1999). Perhaps the clearest indicator of super-adaptive success has been the steady rate of human population increase. Whereas most populations are strictly controlled by available natural resources, it has been possible for humans to far outstrip environmental carrying capacities in nearly every corner of the globe. Yet, explosive population growth has also been the cause of many other problems, such as environmental degradation, resource shortages, widespread disease, etc. (Ehrlich and Ehrlich, 1991).

Long ago, Thomas Malthus (2003) argued that, although humans might be able to outpace nature's carrying capacity in the short-term, in the long run the mathematics of population growth would precipitate disaster. According to Malthus' original predictions, widespread starvation was inevitable due to the fact that food production could only increase arithmetically (i.e., expanding the acreage of arable land one field at a time), whereas population growth was exponential. During Malthus' day, without the advantage of family planning technologies, offspring in typical pre-industrial families could easily outnumber parents by an exponential factor. Consequently, Malthus felt certain that, as populations exploded, demands for food would necessarily exceed available supplies. When that occurred, widespread starvation would ensue with the end result being a catastrophic population crash.

Interestingly, Malthus published his first predictions about overpopulation in the year 1798. At the beginning of the nineteenth century, the global population stood at approximately one billion people. In the years since his prediction, global population has indeed grown exponentially. As of 2011, global population has climbed to about seven billion people. Although it is safe to say that population has mushroomed precisely as Malthus predicted—during the twentieth century alone global population quadrupled from 1.5 billion to six billion people—the global food crisis that Malthus predicted has not occurred. Certainly, there have been persistent and tragic food shortages all over the planet, particularly in the developing world. Nevertheless, the calamitous food crisis that Malthus predicted has not yet transpired. Thus, one must wonder: How have humans avoided such a fate?

Seeing the Future through the Past

It turns out that, like many great thinkers, Malthus attempted to predict the future through the lens of the past. Time is structured such that human experience is always located in the present. In turn, the present can be understood as a dynamic temporal transition point through which time flows toward the future and away from the past. It's as if we are time-surfers; we skim forward on an apparently static foundation in the present while time washes by from the future to the past. Given the one-way flow of time, humans have direct experience with two of the three discernible temporal domains: we occupy the present while preserving memories and other records from the past. Once again, due to the unidirectional flow of time, the temporal dimension with which humans lack direct experience is the future. Never having lived in the future, its specific attributes are largely a mystery. The flow of time would need to reverse in order to acquaint time-travelers with the same level of insight about the future that we currently accumulate about the past and present. Thus, no one can predict the future. At best, we can make educated guesses, based upon extrapolations from the past and present, about how the future might unfold.

Essentially, the future is a process. The fact that the earth has been revolving around the sun for eons is a fairly strong indicator that it will continue doing so in the future. However, as Popper argued, past circumstances, no matter how long they may have persisted, provide no absolute guarantee that similar events will transpire in the future (Popper, 1959). Though the probability is minuscule, an asteroid just might pulverize the earth tomorrow. Thus, the future is a combination of phenomena that give shape to the present blended with dynamics that stimulate change in the future. As such, the future is a construct that is constantly undergoing a process of evolutionary and often unpredictable change. Indeed, one of the most unpredictable instigators of change is the often improbable impact that human agents have upon the structure of unfolding events.

Thus, Malthus gazed into the future through a paradigm that was comprised of eighteenth century expectations. Within the context of the eighteenth century, there was no conceivable means by which to sustain exponentially-increasing populations. As a result, Malthus was convinced that the end was near. Interestingly, in the late eighteenth century, the world as Malthus knew it was about to end, but, importantly, not in the way that Malthus had predicted. Malthus published his prognostications about the presumptive fate of humanity as the age of agriculture was coming to a close. Being unacquainted with the sweeping social, political, economic, and scientific changes that would accompany the Industrial Revolution (Ashton, 1948), Malthus was unable to foresee the innovations that would amplify food supplies sufficiently to keep pace with exploding populations. Technologies such as higher yield grains, fertilizers, pesticides, herbicides, and petroleum-powered machines have generated astounding increases in agricultural yields throughout the industrial era (Pollan, 2006).

Without doubt, the problems that Malthus identified were real. Just as it would have been impossible for NASA to safely land astronauts on the moon using 1950s space technologies, so too would it have been impossible to avert the Malthusian nightmare using seventeenth century agricultural techniques. Exploding populations represented a dire problem, and as Popper argues, life forms must find ways to solve problems, or they will die. In response, humans dealt with the problems associated with population growth and impending food shortages by pursuing an entirely new problematic: industrial society. Having thus dramatically redefined the substance and structure of human society, nineteenth century Europeans set about the process of transforming the social, political, cultural, and technological landscape to make the machine-age a reality.

Mitigating Problems through Progress

In the industrial era, the food supply problems that Malthus foresaw have largely been mitigated. Again, in recent centuries, global population has expanded exponentially, and, in spite of nagging problems with food distribution, the total supply of food has kept pace. Though Malthus would be surprised by this outcome, Popper would not. Again, Popper argued that humans are extraordinarily adept at developing intellectual solutions to survival problems. In successfully identifying such solutions, old problems often become non-issues: in industrially-advanced nations, farmers have been able to produce more food than consumers can eat. In fact, instead of being plagued by shortages, Americans are increasingly plagued by the problem of overabundance (Schlosser, 2001; Spurlock, 2004).

The fact that the population quadrupled during the twentieth century is an undeniable indicator of human problem-solving ingenuity. However, as the population has grown, the degree to which humans have taxed the environment has also increased. For example, our love affair with hydrocarbons (Houghton, 2004) has had a dramatic impact on global climate, including elevating sea levels, shrinking cryosphere, expanding deserts, etc. Thus, successful exploitation of fossil fuels has produced entirely new problems.

Frustrating as this situation may seem, Karl Popper would not be troubled by such developments. Popper argued that, in the game of life, successful solutions to one set of problems invariably generate an entirely new set of problems (Popper, 1992, 1999). When Popper argued that all life is problem solving, he literally meant that survival for every living creature is contingent upon developing workable solutions to environmental problems. Living creatures either develop effective strategies to secure the necessary sustenance, space, and security that they require, or they will expire. Most life forms develop new survival strategies through the genetic evolutionary process. Random genetic mutations that enhance a creature's ability to solve environmental problems (e.g., accessing new food sources, dissuading predators, expanding into new territory, etc.) confer advantages in the struggle for survival. However, as creatures successfully adapt, they inevitably encounter new survival challenges to which they must adapt afresh (e.g., marine mammals successfully reconquered the sea only to encounter hungry sharks in their new environment). Thus, evolution is a never-ending process because every creature must constantly re-adapt to changing survival conditions. Humans are subject to much the same survival pressures as other creatures.

Having taken full advantage of hydrocarbon-age technologies, humans are now confronted with an entirely new set of problems: global warming, ozone depletion, unsecured nukes, pollution, depletion of resources, etc. In spite of our success, the problems that plague humanity seem, if anything, larger and more insoluble than ever. Yet, strange as it may seem, that's actually a good thing.

It is certainly true that, if nothing changes, the problems that humanity currently faces will be irresolvable. Just as Malthus gazed at the burgeoning problems in the eighteenth century and concluded that, for citizens of that era, the situation was hopeless, the same will be true for citizens of the twenty-first century. Einstein summed up this situation thus: "The significant problems we face cannot be solved at the same level of thinking we were at when we created them" (Albert Einstein cited in Fripp, et. al. 2000, p.135). In other words, we can't possibly hope to solve existing problems with existing knowledge. While that might seem to be a hopelessly pessimistic perspective, it is simply a statement of truth—but it is also a call to action. President Kennedy knew that Americans would never be able to solve their problems by malingering in the present. For humans, the key to solving existing problems has always been to create problematics though which to invent entirely new futures.

Indeed, as a result of Kennedy's problematic, the problems that plagued the United States in the early 1960s seem almost comical in hindsight: fear of Soviet expansionism, being on the short end of a strategic technology gap, being at the mercy of the enemies of democracy, etc. JFK inspired the United States to grapple with the most difficult problematic that he could possibly invent. To successfully resolve the problematic, Americans had to aim for the stars. Certainly, along the way, Americans also solved the problem of safely landing astronauts on the moon, but in doing so they accomplished so much more. In striving to achieve JFK's impossible dream, Americans redefined themselves, their nation, and their future.

Americans have had the great fortune to arrive in the twenty-first century as citizens of a powerful, wealthy and technologically-advanced nation. Yet, having solved so many difficult problems in the past, Americans and the entire global community are now faced with even bigger problems. If we have learned anything from the past, it should be that the surest route to a brighter future is to challenge ourselves with the most difficult problematics that we can possible imagine. Though we will never live in a trouble-free world, we can feel safe in the knowledge that, so long as we have the courage to dream of doing the impossible (Kaku, 2008), no problem will ever be too challenging to overcome.

The Next Great Frontier

As America's terrestrial problems escalate, critics have charged that space flight is little more than a futile and expensive hobby. After all, what hope is there that NASA programs will ever resolve practical problems such as winning the war on terror, reversing global warming, reducing the spread of AIDS, or eliminating global hunger? Indeed, space flight is so atmospherically insulated from the real world that NASA programs often appear to be little more than distractions; diverting scarce resources from a plethora of intractable social issues and blasting them into the boundless void of space. In short, what possible benefit can earthlings hope to derive from the billions of dollars required to keep space programs afloat?

History has shown that humans have got all the brains, wherewithal and fortitude to achieve the impossible. We can build a brighter future. All we need are visionary leaders who are prepared to lead the charge toward the Next Great Frontier.

So, why bother with space travel? Because, quite simply, the stars light the way to a brighter future. Space travel paved the way to Kennedy's New Frontier in the 1960s. If the United States remains committed to accomplishing ever greater feats in the future, then we should look to the stars to light our way. Thus, space travel is not a distraction. Space travel represents the path to America's next Great Frontier.

In the process of redefining reality, the boundary between fantasy and reality is often blurred. In the next chapter, I will examine a type of technology that remains more fantasy than reality: artificial intelligence. Is AI the most promising problematic of the twenty-first century? Does AI represent a realistic path to a great new frontier, or is AI simply a distraction: a phenomena borrowed from science fiction that has no real chance to alter the future or lived reality?

Chapter 5
Revolutions in the Making

A Brief Review

Humans are agents of change. As I have argued throughout this book, it is human nature to view reality problematically. In other words, humans are unique among creatures in that they do not treat empirical reality as inevitable. Instead, humans approach reality as a fluid environment that is subject to modification: the shortcomings of a particular environment merely provide creatively-inclined agents with an impetus to modify the state of reality. Thus, humans redefine reality with regularity; however, they do so most effectively when operating within a well-defined problematic. Again, a problematic can be understood as a far-fetched conceptual challenge that can only be achieved through a proactive problem-solving process. While change for the sake of change may appeal to some, knowledge-seekers often do their best work while endeavoring to solve seemingly impossible goals that are coordinated around clearly defined problematics, such as building the great pyramids, inventing heavier than air flight, sending astronauts to the moon, or creating artificial intelligence.

Certainly, some flights of imagination serve no purpose beyond providing a convenient creative outlet, e.g., daydreaming about wonderlands (Carroll, 1897). While such imaginative musings offer evidence of important intellectual creativity, they are not, in themselves, indicative of either problematic knowledge-seeking or human agency. Agency is a specific form of truth-seeking that involves both original thinking as well as a conscious effort to counteract powerfully constraining social influences. I have argued that the clearest expression of agency takes place when individuals redefine reality.

Individuals tend to redefine reality when they encounter facts that are poorly explained by the knowledge systems that dominate their experience—thus, the dissonance that Galileo felt when he spied planets in orbit around Jupiter. Although people often rationalize anomalies, in some cases, anomalies serve as motivators for inventive thinkers to create more truthful explanations for poorly understood facts. I argue that redefining reality is an expression of agency because, in order to make sense of anomalies, it is necessary for individuals to dismantle the influences of social constraints on their thinking. Again, Galileo literally had to throw off the restrictions of Catholic dogma in order to make sense of the anomalies that he had observed. In addition to being an expression of agency, the act of contravening the limitations of prevailing knowledge systems also enables free-thinkers to experience moments of truth.

Moments of truth can be understood as episodes of often exhilarating cognitive clarity that derive from constructing new explanations for previously inexplicable anomalies. Thus, the process of redefining reality is productive of truth in a number of ways. First, by challenging the strictures of established knowledge systems, agents succeed in reducing the intellectual distortions that cloud judgment and inhibit free inquiry. Second, by reducing cognitive distortions, agents are able to evaluate anomalies more thoroughly and, as a result, develop innovative explanations for previously enigmatic phenomena. For example, rather than being cowed by medieval Church dogma, Galileo employed his observations of Jupiter's moons to advance a more adequate heliocentric theory. Thus, moments of truth incorporate both of the essential activities that combine to produce agency: challenging established social structures, and engaging in the kind of original thinking that helps to produce new-and-improved paradigms.

Although anomalies often trigger the process of redefining reality, agents can also redefine reality when they endeavor to resolve problematics. As Popper pointed out, most creatures are limited to solving problems with their biology: when they are genetically ill-fit for an environment, most creatures either randomly evolve genetic solutions or they perish. Humans, on the other hand, are super-adaptable. This means that, rather than depending upon random genetic adaptation, humans solve problems cognitively. Cognitive problem-solving is advantageous because it enables humans to resolve environmental problems with greater speed and specificity than is possible via the biological evolutionary process. In addition, rather than having their fate determined by the whims of environmental change, humans have employed their super-adaptive cognitive skills to modify environments to suit their preferred goals and interests. As a result, by adopting a problematic approach to survival challenges, humans have elevated the process of redefining reality to an entirely different level.

Rather than having their fate determined, as is the case with most creatures, by the resources that are readily available in the environment, super-adaptable humans have created synthetic alternatives to reality. Synthesized realities include, among other things, shelters with artificial lighting, climate control, creature comforts, entertainment and endless supplies of snack food. Within the context of such synthetic environments, human aspirations are bounded less by environmental limits than by the scope of the human imagination. Whereas Popper and others have defined truth as knowledge that is supported by the facts, time and again, humans have concocted problematics that are, at least in their initial stages, utterly unsupported by the facts (e.g., a passion for heavier-than-air flight, conjectures about atomic weaponry, or dreams of flying through the lifeless void of space). After creating a problematic, knowledge-seekers have often, by sheer dint of will, generated the facts that have transformed their flights of imagination into hard and fast truths. Strange as it may seem, imagination is often more important than facts in the process of refining reality.

The Intersection of Fantasy and Reality

Indeed, this is a point that was emphasized by Jean Baudrillard. That is, Baudrillard argued that, in a mass-mediated society, Americans had become particularly adept at concocting elaborate fantasies, or simulations (Baudrillard, 1989). Baudrillard contended that post-industrial Americans vastly prefer consuming—and being consumed by—simulations than living within the confines of the dull, unanimated real world. For Baudrillard, the American obsession with simulations was an indicator of the futility of the postmodern era: when people care more about illusions than reality, the significance of human endeavors dwindles to vaporous insignificance. While I concur with a number of Baudrillard's observations, I differ regarding the utility of simulations. Certainly, many simulations are largely devoid of worthwhile content; television is, in many respects, a vast wasteland (Minow, 1961). Nevertheless, some simulations have helped to inspire the most remarkable feats ever achieved by humankind.

As covered in the previous chapter, the United States' dreams of winning the space race were, in their origins, an outright fantasy. President Kennedy's prediction on May 25, 1961, that the United States would send astronauts to the moon before 1970 was an exercise in science fiction. As of 1961, the United States lacked the technology and anything approaching a workable plan to achieve such a goal. In spite of those shortcomings, Kennedy's lunar landing problematic created an extraordinary opportunity for NASA's super-adaptable community of scientists to redefine the future: Kennedy's problematic stimulated the United States to reinvent itself as the unparalleled leader of the information society (Friedman, 2008).

Rational thinkers who insist that truth must be based upon facts would certainly have had grounds to object. If truth must be supported by facts, then NASA's race to the moon throughout the 1960s was one of the most highly-publicized fantasies ever invented. Yet, for super-adaptable humans, it turns out that truth is not a product of facts alone; what determines the truth in the synthetic landscape of human experience is the combination of imaginative thinking and practical application. Truth is indeed founded upon facts, but not necessarily by the facts that exist at any particular point in time. For super-adaptable humans, truth is often redefined by the facts that, via the magic of problematics, they are inspired to invent.

Thus, it is a fact that on July 20, 1969, Neil Armstrong set foot on the moon. However, Armstrong could never have left his bootprints on the lunar surface if Kennedy had not dared to propose his future-building problematic.

Consequently, redefining reality is a process that generates new truths. In some cases, anomalous facts inspire the redefinition of reality (e.g., Jupiter's moons, ancient fossils, the unaccountable behavior of sub-atomic particles, etc.), while in other cases bold new problematics instigate the invention of entirely new facts (e.g., electronic submarines, machine-powered flight, lunar landings, intelligent computers, etc.). No matter which way the causality flows, social change is generated by human agents who are intent upon eliciting new truths. In either case, truth is a dynamic standard that navigates the boundary between fantasy and reality. In a sense, one could argue that truth can be understood as a description of reality that is supported by the facts. However, one must also recognize that, due to the inventions of ingenious agents, the landscape of the factual universe is constantly in flux. Today's fantasies are tomorrow's facts (Kaku, 2011).

Of course, not every fantasy becomes a fact. Indeed, the vast majority of fantasies will forever remain unrealized dreams: whims that come and go as ephemeral mental constructs without substance or impact on the structure of empirical reality. However, in some cases, it is possible for creative, determined agents to transform their wildly imaginative problematics into things of substance: facts. When fantasies become facts, agents create new truths. In the discovery of those new truths, agents hold the key to transforming reality in the most extraordinary ways.

A Work In Progress

To this point, the redefinitions of reality that I have described have all been a matter of history. From Galileo to JFK, their inspirations have provided an illuminating account of the impact that innovative thinkers often have on the course of history. However, truth-seeking is an ongoing process—something that, I argue, always has been and always will be an essential part of the human makeup. Thus, in what remains of this chapter, I will focus on a problematic that is currently helping to redefine reality in a fantastic new way: the invention of artificial intelligence (AI).

My purpose in examining AI is to emphasize that, when problematics are under development, their ultimate outcome is often far from a matter of certainty. Indeed, as knowledge-seekers endeavor to resolve a problematic, their hoped-for objective can often seem so far-fetched that observers might be inclined to scoff; this was certainly the case with heavier-than-air flight (Tise, 2009), lunar landings (Bizony, 2009), and I daresay it remains the case with artificial intelligence (Nilsson, 2010). Although many people have heard of artificial intelligence, I daresay far fewer are convinced that computers will one day be as smart as humans. Again, if truths are determined by facts, then there is hardly a more elaborate contemporary fantasy than the quest for artificial intelligence.

Another reason that I will examine the AI problematic is to emphasize that truth-seeking is a necessarily contentious process; agents must challenge and transform both the structure of knowledge and the nature of reality in order to elicit new truths. For example, in spite of JFK's determination, a pre-1970 moon landing was never a certainty until the very moment that Armstrong hopped off the LEM's ladder. Yet, the implausibility of Kennedy's dream has been moderated somewhat by the fact that such a seemingly impossible feat has become a matter of historical certainty. Thus, to gain a better appreciation for the psychosocial turmoil that accompanies the problematic truth-seeking process, I would like to examine a reality-transformation initiative that is currently underway.

Many scientists, scholars, technologists and other dream-weavers have been laboring for decades to transform the fantasy of AI into a reality (McCorduck, 2004). During the past half century there have been monumental advances in the field of AI research, however, in some respects, the entire project still remains in its infancy. That is, the computer on which I am currently composing this passage is no more capable of intelligent conversation than the PC that I purchased in 1990. Certainly, computer technology has advanced enormously over the past twenty years, however, for most users, intelligent computers remain little more than a figment of over-active imaginations—and that is precisely what makes AI such an instructive example at this point in the discussion.

Artificial intelligence is a problematic in search of the facts that might one day make it a reality. Thus, rather than being the subject of mild historical interest, AI remains an actively contested problematic. Indeed, there are perfectly valid reasons to doubt that researchers might ever create AI (Searle, 1980).

Computers do not think for themselves, and it requires a substantial leap of faith to believe that such hopelessly inarticulate devices ever might. Indeed, in the discussion that follows, I will examine a variety of issues which illustrate that a healthy skepticism of the entire AI project is warranted. As with most problematics, doubters and dissenters will appear more rational—their feet more firmly planted on terra firma—until the moment that the fantasy they forswear is realized. If and when it becomes possible to create AI, the world will change dramatically. As with circumnavigation (Bergreen, 2003), aviation (Tobin, 2004) and space travel (Erickson, 2005), such a redefinition of reality will propel both doubters and believers into utterly new territory. It is difficult to imagine what living in a world with truly intelligent machines might be like. Such perplexity is reasonable because, just as pre-industrial farmers could not anticipate the sweeping changes that would accompany the industrial revolution, intelligent machines will certainly introduce vast and unforeseeable changes into a radically redefined reality. Indeed, from the context of that new reality, were it ever to exist, AI skeptics will no longer seem more rational than those who dare to dream. Instead, AI skeptics will seem as quaint and unimaginative as those who were once convinced that the earth was flat.

AI: A Contemporary Problematic

Dramatic improvements in computing capacity have been commonplace throughout the information age. In 1965, Gordon Moore predicted that integrated circuit capacity would continue to double every two years until well into the foreseeable future (Brock, 2006). In the years since his observation, Moore's Law has predicted computer chip progress with astounding consistency. Year after year, processors have become smaller and faster, while information technologies have become increasingly pervasive and portable. Yet, sophisticated as handheld wireless technologies have become, they are primitive in comparison to the technologies needed to create artificial intelligence.

Artificial intelligence represents a threshold in computing that will transform machines into human-like intellectual entities. Indeed, rather than merely imitating human intelligence, some AI researchers have expressed a desire to create "a new form of life" (See the homepage of AI Research, www.a-i.com, for just such a declaration).

In a world grown used to rapid technological progress, it hardly stretches the imagination to contemplate next-generation computers that are substantially smaller and more powerful than their predecessors. However, the goal of transforming computers into sentient beings is an entirely different matter. In spite of the apparent sincerity of the AI Research initiative, their stated goal of "creating a new form of life" is a somewhat dubious objective. To begin with, I must quibble about the distinctions between life and intelligence. There are many creatures that are, technically speaking, alive, but that are also entirely lacking in intelligence. For example, amoebas are alive. However, being unicellular life forms, amoebas are devoid of the complex neurological structures required to produce intelligent thought. Thus, life is not contingent upon intelligence. On the other hand, intellectual operations, no matter how sophisticated, do not constitute life. Life is a characteristic of creatures that are imbued with self-sustaining biological functions. Should computers ever reach the point of exhibiting self-directed intelligence, the question of whether or not those machines ought to be considered "alive" will probably become a topic of vigorous debate. For now, the distinction between living organisms and smart machines is both chasmic and self-evident.

Yet, long before we confront the conundrum of living machines, we must first consider the more immediate plausibility of artificial intelligence. In other words, is it possible to simulate that most distinctive of human traits, and install it in machines? Certainly, computers are constructed by humans and, as a result, they are indelibly imprinted by the intelligence of their creators (Carter, 2007). In addition, computer output has a tendency to mimic human intelligence. Humans communicate with language, both semantic and numeric, and, consequently, so do computers (Brighton and Selina, 2003). Still, it is one matter to suggest that computers represent a reflection of human intelligence—which they most assuredly do—but it is an entirely different matter to assert that computers might one day think intelligently.

Intelligence, though common among humans, is nevertheless, a rare, complex and imperfectly understood phenomenon. Routinely as we may use the term in reference to human brain function, a precise definition of intelligence remains elusive. Working definitions abound (e.g., the ability to acquire, create and apply knowledge), but none seem wholly satisfactory. In addition, although it is clear that human intelligence is rooted fundamentally in complex neurological activities, it is not clear precisely what combination of brain functions generate the singular experience of sentience. Brains alone, no matter how large, do not produce intelligence. Intelligence only emerges in humans who have fully-formed brains and who are also the beneficiaries of long-term socialization experiences (Mead, 1934; Freud, 1961; Minsky, 1985, 2006; Piaget, 1951). Given all of the variables associated with intelligence, no formula or definition has yet emerged that fully captures the complexity of intelligence in humans. Lacking a complete understanding of this most essential human trait, the chances of capably synthesizing an artificial form of intelligence seem less likely still.

Yet, challenging as such an endeavor may be, there are numerous public, private and non-profit institutions that have maintained ongoing AI research programs for decades. AI researchers have made impressive progress in the endeavor to create more sophisticated machines. However, the more that AI researchers have learned about constructing smarter machines, the more daunting the prospect of developing a full-fledged form of machine intelligence appears. With each step down that particular path, the horizon seems to shrink even further into the distance. Indeed, Marvin Minsky, a pioneer in the field of AI research, has defined intelligence in the following terms: "*Our minds contain processes that enable us to solve problems we consider difficult. 'Intelligence' is our name for whichever of those processes we don't yet understand*" (Minsky, 1985, p. 71, italics in original). Nevertheless, AI research is one of the most beguiling technological challenges that humans have ever undertaken. Perhaps because it remains such a seemingly impossible objective, researchers remain irresistibly committed to transforming this ambitious problematic into a reality.

The Turing Test

In 1950, Alan Turing published a seminal article on machine intelligence (Turing, 1950). In the article, Turing proposed a test to determine the point at which machines would achieve a definable level of intelligence. Turing's test involved a remote, three-way conversation between two humans and one machine. In the context of the test, one of the humans, the interrogator, would communicate via an electronic media link with the two interviewees; one being human and the other being a computer. It would, thus, be the interrogator's job to determine which of the two interviewees was a machine. Turing argued that when it became impossible for an interrogator to reliably distinguish between their human and machine interactants, the conversant computer in question would have passed the Turing Test. In other words, Turing argued that when machines become communicatively indistinguishable from humans, those machines will have become artificially intelligent.

Since 1950, the Turing Test has inspired a great deal of debate concerning the prospect of intelligent machines. In Turing's day, computers existed, but they were far too primitive to exhibit properties of intelligence. Nevertheless, Turing believed that by the turn of the twentieth century, computers would advance to the point of exhibiting human-like intelligence.

Whether referring to humans or machines, Turing believed that three key components combined to generate displays of intelligence:

1. Memory
2. Processing power
3. Programming

Although Turing acknowledged that the human brain was an electronic nerve center, he argued that electricity was not a fundamental requirement for machine intelligence. This is the case because Charles Babbage successfully designed a mechanical computation device in the nineteenth century (Swade, 2001). Although unlikely, Turing argued that it was theoretically possible to develop a mechanical computational device that could achieve machine intelligence. Thus, Turing asserted that, contingent upon sufficient advancements in computer memory, processing power and programming, researchers would soon be able to create an age of intelligent machines (Kurzweil, 1992).

Arguably, Turing's speculations about AI were more academic than JFK's plan to land astronauts on the moon. Nevertheless, each of their grand visions is clearly identifiable as a problematic. Once again, a problematic can be understood as an exceedingly challenging objective that requires knowledge-seekers to invent new facts in order to achieve a specified goal. Different as they may have been, nevertheless, there are important similarities between Turing's and Kennedy's problematics. Both advocated a vision of the future that included elements of pure fantasy. Also, each created their problematic for the express purpose of motivating the scientific community to transform their respective fantasies into reality.

Although Kennedy's call to action came later, JFK had the advantage of being a United States president. Thus, being able to summon the resources of an entire nation, JFK blazed a trail to his new frontier on an astoundingly brief timetable. On the other hand, even though he may have been a brilliant and influential scientist, Turing could only focus a fraction of such attention on his problematic. As a result, even as we enter the second decade of the new millennium, the notion of artificial intelligence remains more fantasy than reality. Yet, although the AI problematic has progressed somewhat more deliberately, there is evidence that technology is taking gradual, but steady steps toward realizing Turing's dream.

Rising to Turing's Challenge

For example, Turing argued that one way to engineer increasingly intelligent machines would be to design computers to play games (Turing, 1950). Consistent with Turing's proposal, IBM developed Deep Blue, a supercomputer that was designed to play championship-level chess. In 1996, Deep Blue earned the notoriety of becoming the first computer to defeat a reigning world champion, Garry Kasparov, in a single chess game (Newborn, 2003). Although Deep Blue eventually lost the match, that near-miss only inspired IBM researchers to redouble their efforts.

In 1997, Kasparov met Deep Blue for a rematch. During the intervening year, IBM developers incorporated significant upgrades into their computer, which inspired a modified, but unofficial nickname for the computer, *Deeper* Blue. Employing a massively parallel, brute force architecture, the 1997 version of Deep Blue was capable of evaluating two hundred million chess positions per second. This was twice as fast as Deep Blue had been in 1996. Further, Deep Blue '97 could also search, on average, six to eight moves ahead in a particular chess game. However, the computer also possessed the more extraordinary ability to search as far as twenty moves ahead. Relying on its brute force computing capabilities, Deeper Blue succeeded in defeating Garry Kasparov. Impressive as that feat may have been, there is some debate about whether Deep Blue's achievement in 1997 was truly a demonstration of a new threshold in machine intelligence.

For his part, Garry Kasparov conceded that he saw evidence of profound intelligence in Deep Blue's moves. However, rather than congratulating Deep Blue, Kasparov contended that his opponent's chess strategy could only have been orchestrated by human intelligence. Essentially, Kasparov claimed that he had been cheated; alleging that humans had interfered with Deep Blue during the course of individual games (Hoffman, 2007, p. 114). If true, such interference would have constituted a violation of the match rules, and it would also have contravened the spirit of the competition, which presumably pitted human against artificial intelligence. The IBM team dismissed Kasparov's accusation by insisting that they had only made modifications to Deep Blue during the intervals between games, which, by the way, was permitted by match rules. Thus, IBM emphasized that, true to the spirit of the competition, during the course of individual chess games, Deep Blue's decisions were entirely its own.

To some extent, the situation remains unresolved. When Garry Kasparov originally requested access to Deep Blue's log files, IBM refused. It is worth noting that IBM subsequently acceded to the request and posted copies of the log files on the Internet (www.research.ibm.com/deepblue/watch/html/c.html). Curiously, following Deep Blue's unprecedented success, IBM made the somewhat puzzling decision to dismantle the computer. Thus, lingering questions about the veracity of Deep Blue's performance during the historic 1997 chess match will remain forever unresolved.

Setting aside the controversy, it is clear that, in Deep Blue, IBM managed to advance the state of computer technology to a spectacular new level. Not only did IBM build the world's most powerful chess-playing machine, but IBM also identified the computing level at which machines could outplay even the most skilled human opponents. Indeed, with Deep Blue, IBM arguably took computing a step beyond the Turing threshold. That is, Turing argued that machine intelligence was contingent on convincing interactants that machines were roughly equal in intelligence to humans. Whereas, in Deep Blue, IBM created a machine that *exceeded* the skill of the reigning world chess champion. In doing so, IBM surmounted a major technological threshold, but is it fair to say that in defeating Kasparov, IBM also created a new form of intelligence?

Would Deep Blue Pass the Turing Test?

Extraordinary as Deep Blue's achievements may have been, the computer was an exceedingly unidimensional machine. Outside the confines of chess, Deep Blue lacked even the most trivial forms of intellectual versatility, e.g., commenting on the prevailing weather, griping about the rigors of business travel, agonizing over the pressures of championship chess competitions, etc. Within the narrow domain of chess, Deep Blue's computational power was superhuman; however, outside of those parameters, Deep Blue was non-starter. Thus, with regard to the Turing problematic, IBM had succeeded in advancing the state of computer technology to an entirely new level, however, IBM still fell far short of creating a full-fledged form of machine intelligence.

Intelligence is a phenomenon that is of particular importance to humans. Arguably, intelligence is the quality that distinguishes humans from all other creatures and, whether rightly or wrongly, confers upon humans a unique sense of identity and superiority. Rene Decartes' (1901) famous statement, "Cogito ergo sum," captures the essence of this special quality. Sentience, or the cognitive faculty to experience self-presence, "I think, therefore I am," has equipped humans with a privileged sense of purpose in the cosmos.

Thus, one could argue that it is both fitting and somewhat ironic that scientists are now endeavoring to simulate that most essential of human traits, intelligence, and install it in machines. Certainly, the prospect of imbuing machines with intelligence remains, at this juncture, a problematic fantasy. Computers have certainly become much more sophisticated in the decades since Turing published his seminal article. In addition, computers have become more inextricably embedded in the contemporary cultural landscape (Dretzin, 2010). Yet, sophisticated as computers have become, there are none, as yet, that have a snowball's chance of passing the Turing Test—and that includes IBM's latest AI initiative, Watson (www.watson.ibm.com/index.shtml) (Thompson, 2010).

Again, Turing argued that three key computing features were essential to generate the appearance of intelligence in machines: memory, processing power, and software. However, by now it appears as though intelligence is comprised of more than those simple ingredients. That is, just as Turing predicted, memory, processing power, and software have advanced by leaps and bounds since 1950. However, in spite of those extraordinary refinements, we have not yet witnessed a cognitive revolution in machine intelligence. Precisely what sort of psycho-social ingredients combine to produce human intelligence remains an abiding mystery. Indeed, before we can obtain a better understanding of the biological and cognitive apparatuses that are required to produce intelligence, it is important to know exactly what we are referring to when we use the term intelligence.

The Hazards of Operationalization

Scientists have long wondered how to define intelligence (Gould, 1996). As with any concept, it is nigh on impossible to establish an exacting definition of intangible sensory phenomena. For example, despite the best efforts of poets, philosophers and truth-seekers, humanity is no closer to a precise definition of love than we were millennia ago. For example, in *The Emotion Machine*, Marvin Minsky (2006, p. 9-12) takes a stab at defining love, but in the end concedes that a precise definition is not possible because the term seems to mean something different to each person who uses it: "Returning to the meanings of Love, one thing seems common to all those conditions: *each leads us to think in different ways* (2006, p. 12, emphasis in original).

The difficulties that confound definitions of love similarly stymie attempts to define intelligence. Quite simply, the word intelligence means different things to different people. For example, in the nineteenth century, the field of craniometry emerged with the ostensible goal of developing more exacting measures of intelligence. Having established a link between brain function and cognition, craniometrists proceeded on the assumption that there must also be a direct connection between brain size and intelligence. In other words, if intelligence is a product of brain function, then it stood to reason that large-brained people would be more intelligent than small-brained people. Thus, early craniometrists operationalized intelligence as a function of cranial volume: greater cranial volume = higher intelligence.

Using a variety of packing materials, from mustard seed to lead shot, craniometrists conducted extensive measurements of cranial volume on skulls from human populations all over the world. Unfortunately, as Gould explains, craniometrists transfused their studies with heavy-handed ethnocentric biases (Gould, 1996, p. 88-89). In other words, craniometrists often embarked on their studies with the explicit intention of demonstrating that the people who happened to be most like themselves (i.e., same gender, race, ethnicity) were also graced with the greatest quantities of intelligence. Thus, European craniometrists were determined to show that they and their brethren were smarter than everyone else. Oddly, and not very intelligently, in many craniomentric studies, skin color often became the weightiest factor in determining "intelligence" (Gould, 1996, p. 97).

Yet, committed as they were to reconfirming racial bias in their studies, more than one craniometrist encountered evidence that challenged their orienting assumptions (Gould, 1996, p. 100-101). Undeterred, craniometrists simply developed alternate analytical lenses that tidily preserved their racism while minimizing the complications of countervailing evidence (Gould, 1996, p. 119).

In the end, Gould argues that craniometrists fumbled their analyses of intelligence in two crucial ways. To begin with, for craniometrists, ethnocentrism trumped science. Rather than permitting themselves to learn from their data, craniometrists routinely distorted evidence that failed to agree with their preferred conclusions. Thus, craniometrists violated the most fundamental guiding principles of good science and, in so doing, invalidated their studies.

Although difficult to believe, Gould asserts that there is an even deeper flaw in early craniometric studies. Craniometrists failed to establish that there was, in fact, any relationship between cranial volume and intelligence. Again, craniometrists embarked on their studies with the false, but unassailable assumption that European males were more intelligent than everyone else (Gould, 1996, p. 116). Proceeding from that assumption, craniometrists manufactured results that papered over every other consideration. In other words, craniometrists believed that their skull measurements represented a precise measure of an incontrovertible natural phenomena: big brains rule.

Certainly, the correlation between brain size and intelligence tidily explains intellectual differences between humans and smaller-brained creatures. For example, animals such as birds, mice and monkeys, are endowed with smaller brains, and, consequently, their reasoning abilities are comparatively less impressive: humans have constructed great civilizations, whereas birds, mice and monkeys haven't even invented the wheel. Unfortunately, the correlation between brain size and intelligence breaks down when we consider larger animals, such as elephants and whales. Of course, the solution to that particular conundrum lies in factoring another ratio into the equation: brain size vs. body mass. Thus, whales and elephants may have larger brains than humans, but their bodies are also considerably more massive (Freeman and Kreuter, 1994, p. 4). When controlling for body size, humans are comparatively much brainier and, therefore, exhibit relatively greater intelligence than other large-brained mammals.

That said, it is important to recognize that the above conclusion is at least partially a product of conceit: the types of intelligence that humans esteem most highly are those that, no surprise, resemble human intelligence most closely. Thus, who is to say that large-brained whales are in an absolute sense less intelligent than humans? Perhaps, by maintaining a serene balance with their natural environs, whales are in fact demonstrating the highest form of intelligence. Creatures of so-called inferior intellect may one day have the last laugh. Nevertheless, for now, because whales and other creatures have failed to impress humans with their intellect, humans generally leap to the conclusion that nonhumans are not intelligent. Thus, before we examine the concept of intelligence any further, it is important to acknowledge that humans define intelligence egotistically: intelligence is that which intelligent *humans* do.

g: Intelligence as a Single Number

Gould (1996) argues that, thus far, every attempt to measure intelligence has been an abject failure. Whereas craniometrists failed because of their explicit ethnocentric biases, Gould argues that intelligence testers, or psychometricians (Herrnstein and Murray, 1994), have also failed for similar reasons. In essence, Gould argues that, rather than measuring intelligence, IQ tests measure proficiency in a particular cultural context. In their earliest days, intelligence tests were employed as a means to prevent mental defectives (i.e., terrified, illiterate, non-English-speaking immigrants) from entering the United States (Gould, 1996, p. 195). Even at the close of the twentieth century, Gould believes that intelligence tests remain more a gauge of cultural immersion than a measure of intelligence.

Indeed, Herrnstein and Murray acknowledge that many people doubt intelligence can be meaningfully represented by IQ test scores. Yet, in spite of those doubts, Herrnstein and Murray simply assert that intelligence tests capably measure intelligence, "IQ scores match, to a first degree, whatever it is that people mean when they use the word *intelligent* or *smart* in ordinary language" (Herrnstein and Murray, 1994, p., 22. Italics in original). Still, Herrnstein and Murray caution against the idea that intelligence can be gauged by making casual observations.

> One of the most insidious but also widespread errors regarding IQ, especially among people who have high IQs, is the assumption that another person's intelligence can be inferred from casual interactions. Many people conclude that if they see someone who is sensitive, humorous, and talks fluently, the person must surely have an above-average IQ...This identification of IQ with attractive human qualities in general is unfortunate and wrong (Herrnstein and Murray, 1994, p. 22).

Intelligence, Herrnstein and Murray insist, is a phenomena that can only be capably measured by IQ tests. Nevertheless, having made that pronouncement (and also in violation of their own admonition), Herrnstein and Murray also make the contradictory assertion that intelligence can be understood as a quality of intellectual agility that people tend to recognize in others.

> . . . high intelligence has earmarks that correspond to a first approximation to the commonly understood meaning of smart. In our experience, people do not use smart to mean (necessarily) that a person is prudent or knowledgeable but rather to refer to qualities of mental quickness and complexity that do in fact

> show up in high test scores. In everyday language, people who are called very smart tend to have high IQs (Herrnstein and Murray, 1994, p. 22).

Thus, g, or IQ is roughly equivalent to observable displays of intelligence.

Interestingly, Herrnstein and Murray end up defining intelligence in much the same terms as Alan Turing. Of course, we must keep in mind that Turing was referring to machine intelligence, whereas Herrnstein and Murray claim to have isolated a valid measure of human intelligence. Nevertheless, in each case, Turing and Herrnstein and Murray associate intelligence with characteristics that interactants display in order to convince others that they are intelligent. Precisely what intelligence means beyond an individual's ability to appear smart remains unspecified. With such an unsatisfactory definition as a departure point, it is easy to understand why many people object to Herrnstein and Murray's broader assertions about intelligence (Jacoby and Glauberman, 1995; Kincheloe, et. al. 1997).

Intelligence as *Savoir Faire*

It is difficult to make convincing progress in any field if one's foundational definitions are deficient. Just as people object to Herrnstein and Murray's suggestion that intelligence can be equated with displays of smartness (Fraser, 1995), there is reason to doubt Turing's contention that machine intelligence can be gauged by a computer's conversational skills. Savoir faire is far from a perfect measure of intelligence. For example, in 1966, Joseph Weizenbaum (1976) composed a computer program, ELIZA, that was designed to create the illusion of AI by having a distinctly unenlightened computer play the role of a Rogerian psychotherapist. That is, ELIZA conversed by doing nothing more than rephrasing comments posed by human interactants:

> Human: I feel miserable today.
> ELIZA: Do you often feel miserable today?

Weizenbaum readily admitted that ELIZA was only intended to create the appearance of intelligence—and, thus, take a credit-worthy swipe at passing the Turing Test—while sidestepping "the problem of giving the program a data base of real-world knowledge" (Weizenbaum 1976, p. 188-189). Thus, Weizenbaum succeeded in creating an illusion of machine intelligence, but one that was entirely devoid of substance. Interestingly, even though Weizenbaum was forthcoming about ELIZA's simplistic capabilities, users would often take ELIZA's feedback seriously. Clearly, such earnest user responses speak to the benefits of Rogerian psychological therapies, but they also identify a critical flaw in the Turing Test: intelligence cannot be adequately gauged by conversational skills alone, particularly if a computer has been cleverly designed to deceive its human interactants. As Herrnstein and Murray warned, there are dangers in treating conversational skills as measures of intelligence. Of course, this makes one wonder afresh why Herrnstein and Murray would, in their own turn, correlate intelligence with conversational displays by apparently smart people (1994, p. 22).

Even though ELIZA lacked any capacity for intelligent exchanges, ELIZA's conversational skills were sufficiently fluid to convince users that ELIZA was a sentient, even a professionally skilled, therapeutic interactant. Thus, one must wonder if IQ tests actually measure intelligence or merely test-takers' ability to appear intelligent. As ELIZA illustrated, there is a big difference between those two things, i.e., just because a parrot can speak does not mean the parrot knows what it's talking about.

Nature vs. Nurture

In his text, *The Society of Mind*, Marvin Minsky (1985), in agreement with Stephen Gould, argues that it is unwise to construct overly simplistic operational definitions of intelligence. Minsky believes that, like love, intelligence is a suitcase term that incorporates so many different meanings, no single definition can do the term justice:

> it isn't wise to treat an old, vague word like "intelligence" as though it must define any definite thing . . . the very concept of intelligence is like a stage magician's trick. Like the concept of *"the unexplored regions of Africa,"* it disappears as soon as we discover it (Minsky, 1985, p. 71, italics in original).

Thus, Minsky believes that it has been impossible to develop a singular definition of intelligence because intelligence is not a singular conceptual construct. Rather, intelligence is a compendium of innumerable lower-order operations. Further, intelligence is also a concept that changes every time we learn new things about the thinking process (Minsky, 2006, p. 109-111). Therefore, from Minsky's perspective, understanding intelligence is not contingent upon constructing a precise conceptual definition. Instead, Minsky believes that, to understand intelligence, it is essential to develop an awareness of the multifarious cognitive processes that combine to produce what we ultimately perceive as intelligence. Thus, we use the term intelligence as a shorthand to bracket a phenomenon that is too complex to adequately describe otherwise.

In addition to recognizing that intelligence is dependent upon the complex interaction of innumerable neurological structures (i.e., nature), Minsky also acknowledges that intelligence only emerges as a result of long-term socialization processes (i.e., nurture). Psychologists, such as Freud (1961) and Piaget (1951)—not to mention sociologists, such as Mead (1934), Cooley (1902) and Blumer (1969)—have argued that, though humans may be born with brains, they only develop well-functioning minds through a process of extended social interaction. Again, Descartes' statement, "I think, therefore I am," implies that sentience is contingent upon the ability to reflect upon oneself. Sociologists have long maintained that individuals cannot develop a mind in the absence of interaction with others. For example, children who lack intensive early childhood socialization experiences do not develop identities or fully-formed cognitive functions (Davis, 1940). Thus, Minsky argues that computational power alone, no matter how large or fast an information processor may be, is not sufficient to produce intelligence. Intelligence in humans and, by extension, artificial intelligence in machines can only be created through an elaborate process of social intellectual development (Minsky, 2006).

Thus, social interaction is crucial to the constitution of self, mind and intelligence. In addition, the only way to demonstrate intellectual competence is through communication. In other words, although an inarticulate individual might be a genius, without communication, there is no way to prove that such genius exists.

Take, for example, the extraordinary case of Stephen Hawking, who argues that he became a more focused intellectual achiever as a case of Lou Gehrig's disease robbed him of his physiological capabilities (Larsen, 2005). Indeed, it was only as a result of the arrest of his disease, combined with the assistance of special computer technologies, that Hawking has been able to communicate his intellectual achievements. Thus, Hawking is a genius, and, arguably, he would be a genius whether he could communicate or not. However, if the technology did not exist to permit Hawking to communicate his ideas, no one but Hawking would ever have known of his intellectual exploits in theoretical physics. Thus, this reemphasizes Turing's original point that intelligence tests, whether human or artificial, must be predicated upon communicative competence.

Franken-telligence

As a means of equipping a machine with the necessary communicative competencies to pass his intelligence test, Turing proposed the idea of building a computer, "with the best sense organs that money can buy, and then teach it to understand and speak English" (Turing, 1950). Thus, much like Dr. Frankenstein, Turing speculated that it might be possible to generate AI spontaneously by cobbling together a set of human-like sense organs and then animating them with a jolt of life-giving electricity. With such a proposal in mind, I think it is important to pose the question: At what point does the human mind become intelligent?

Although healthy, fully-formed infants have the potential for intelligence, at the moment of their birth—and though proud parents might protest—one could not accurately describe infants as intelligent. Certainly, infants are bursting with potential and it only takes a few years of love and devotion to activate their intellectual gifts, but at the outset, infants are pre-intelligent beings at best. Thus, it is somewhere along the way in life's journey that humans transition from potentially bright creatures to verifiably intelligent human beings. Absorbing vast quantities of information through various sensory apparatuses is arguably an essential precursor to intelligence, however, information acquisition alone does not produce intelligence. That is, libraries acquire enormous quantities of information, however, it makes no sense to describe libraries as intelligent. Intelligence involves more than information acquisition; intelligence is also contingent upon an ability to manipulate information. Building upon this point, Minsky argues that geniuses are simply people who are particularly adept at organizing and applying what they learn (2006, p. 277). "It's not enough to learn a lot; one also has to *manage* what one learns" (Minsky, 1985, p. 80. Italics in original).

How do infant humans transition from being relatively passive information-absorbers to active creators of information? I would hazard that, if someone were to find an answer to that question, then that person would also have solved the fundamental enigma of intelligence. Certainly, the ability to manipulate information is associated with the aptitude that humans require to interact with others and, ultimately, to reflect upon themselves. However, there is much concerning the mechanics and, more importantly, the aesthetics of intelligence that remains a mystery—and will probably remain so for a long time to come. Thus, I believe that there is some truth to the notion that, though humans are intelligent enough to experience intelligence, humans are not intelligent enough to understand intelligence (Dreyfus, 1992; Hofstadter, 1979). At least, not yet.

A Working Compromise—Perhaps

Certainly, many AI researchers would object to such an assertion. Raymond Kurzweil echoes Marvin Minsky's sentiments by suggesting that the definition of intelligence tends to be a "moving frontier" (Kurzweil, 1992, p.14). Minsky's point is that definitions of intelligence tend to change as researchers learn more about intelligence. In turn, Kurzweil argues that every time there is a major advance in the field of artificial intelligence, rather than treating the achievement as a major step forward, Kurzweil contends that perceptions about what qualifies as AI become more stringent: "artificial intelligence is the study of computer problems that have not yet been solved" (Kurzweil, 1992, p. 14).

I think it is fair to say that, since Turing published his AI problematic in 1950, advancements in the field of information technology have been extraordinary. In addition, I believe that Minsky and Kurzweil are correct in stating that, due to such advancements, public attitudes toward information technology have changed. Indeed, it has been my thesis throughout this book that humans have a unique ability to modify their beliefs when confronted with new facts, and the pursuit of AI is an instructive example of that phenomenon in action.

As often occurs in the process of resolving problematics, Turing's challenge inspired IT developers to invent one generation of facts after another. Whereas in the mid-20th century, computers were vast, mysterious and inaccessible, by the turn of the 21st century computers had become personal, portable and an integral feature of everyday reality. Thus, in the face of those changes, it stands to reason that perceptions of computers and the role that they play in shaping the present and future would also change. In 1982, Time Magazine captured the burgeoning sense of wonder inspired by the PC revolution by naming the personal computer its Man of the Year. It only added to the sense of historical moment that this was the first time an object, instead of a famous man or woman, was accorded this signal honor.

In the years since, computer technology has advanced by leaps and bounds, but, having grown used to ground-breaking technological achievements, the public has come to view IT innovation with a certain degree of complacency. No matter how wondrous the latest computer advancements might be, we expect today's technology to be outdated in a twinkling. Thus, there are grounds to support Kurzweil's contention that, as technologies have become more sophisticated, the public has also become increasingly difficult to impress. Kurzweil's point is that AI is not as far-fetched a fantasy as many people seem to think. Indeed, not only does Kurzweil believe that AI will inexorably evolve out of human intellectual capabilities, but he believes that, were we to properly acknowledge the exceptional IT developments that have already taken place, the public would rightfully conclude that AI already exists.

> When the artificial intelligence field was first named at a now famous conference held in 1956 at Dartmouth College, programs that could play chess or checkers or manipulate equations, even at crude levels of performance, were very much in the mainstream of AI . . . we no longer consider such game-playing programs to be prime examples of AI, although perhaps we should (Kurzweil, 1992, p. 15).

Thus, Kurzweil believes that it is only due to an escalating threshold of expectations that the public has failed to acknowledge that AI already exists.

I agree that public perceptions about information technology have changed dramatically. In the 1950s, the idea that people would one day have computers in their homes and offices (much less in their pockets!) was pure science fiction. That said, I do not agree with Kurzweil's contention that that the threshold for defining AI has been entirely fluid. Since 1950, the Turing benchmark to identify an artificially-intelligent machine has not changed one bit. Earlier, I pointed out that the QWERTY computer on which I am typing had no more ability to engage in a fluid, human-like conversation than the PC clone that I purchased two decades ago. My point was that, keeping the Turing-articulate threshold in mind, my iPad tablet is no more of an enlightened chatterbox than my old desktop PC. Though computer technology has certainly changed, the distinction between AI and non-AI technologies has remained static. Though we might wish it were otherwise, computer developers have a long way to go before they'll be able to create anything close to a Turing-articulate machine.

The Science of Unintended Consequences

In science fiction, AI is often presented as a *fait accompli*. In *2001: A Space Odyssey*, and old *Star Trek* episodes, Turing-articulate computers just exist. Viewers need not concern themselves with how such extraordinary machines came to be. They simply exist. After all, it being the future and computer technology bounding ahead the way it does, it is difficult to imagine a future that is absent of some version of Turing-articulate computers. Thus, it can be tempting to believe, as Kurzweil contends, that AI is inexorable.

Among AI enthusiasts, Kurzweil stands out as one of the most optimistic of an enterprising group of problem-solvers. For Kurzweil, AI is not a fantasy, it is a reality. Part of Kurzweil's confidence derives from the fact that he has already created a variety of proto-AI technologies (Kurzweil, 1999, p. 84-85). Yet, more than that, Kurzweil is convinced that AI is logically inexorable: Kurzweil argues that, the very existence of human intelligence created the essential preconditions for the eventual usurpation of human intelligence by a superior form of machine intelligence (Kurzweil, 2005).

For a variety of reasons, I am less convinced that a HAL-like version of AI will inexorably emerge. As I explained earlier, it is never possible to see the future clearly through the distorted lens of the present; the future will only be intelligible through the paradigms that humans invent to understand the as-yet-unknown future that they will help to create. Nevertheless, I suspect, in agreement with Kurzweil, that information technologies will advance by intensifying the cyborg-like (Clark, 2003) fusion between humans and IT. In other words, people have already begun plugging computers into their ears. Advancing technologies will likely involve adorning ourselves with smaller, more powerful IT devices—or perhaps even installing them in our bodies (Hsu, 2010). And, when we arrive at the point of installing computers in our bodies, in what strict sense will human cognition be entirely distinguishable from machine-based intelligence? Intelligence is an experience that is rooted in human biology and sociality and that also happens to be enhanced through information technologies. That has certainly been true ever since humans invented the printing press (Briggs and Burke, 2009) and the same will continue to be true as information technologies become even more sophisticated.

Still, for the sake of argument, let's say that in the years ahead, researchers finally manage to transcend the Turing threshold. At that point scientists will be able to celebrate a landmark achievement in technological development: once again, knowledge-seekers will have resolved a seemingly impossible problematic. However, in doing so what else will they have accomplished?

The Solution to End All Solutions—or Not?

In the United States, the 1970s were a decade of pessimism. Since the conclusion of WWII, the United States had grown used to playing the role of superpower. An important reason for United States ascendancy was the predominant position that the United States came to occupy in the global industrial economy. During WWII, the United States benefited from the fact that, as opposed to other combatant nations, wartime destruction strengthened rather than demolished America's infrastructure. While bombs dropped in Europe and Asia, American factories throttled up from a pre-war Depression into wartime overdrive. Following WWII, the United States economy was stronger than ever.

Though plagued by cold war anxieties, the United States economy boomed all the way through the 1950s and 1960s. During those decades, the United States produced more and better industrial goods than any other nation. Times were so good that it seemed like the United States would remain the world's industrial leader until far into the foreseeable future. However, as the United States entered the 1970s, things suddenly changed.

Partly due to short-sighted planned obsolescence strategies, the United States' industrial dominance went into decline during the 1970s. International competitors in the automobile industry, the heart and soul of industrial America, began gaining ground. The unkindest cut came when Asian and European competitors began horning in on America's formerly inviolable domestic car market. As cheaper and higher quality industrial products began flooding into the United States, America's industrial infrastructure started to crumble (Bluestone and Harrison, 1982). Industries that had once thrived in the United States began moving overseas. The once prosperous industrial belt rapidly disintegrated into a sprawling rust belt. Mighty as industrial America had once been, it appeared as though America's days as a global leader had already gone by.

Certainly, during the 1970s-80s there was a lot of doom and gloom in the United States. Yet, instead of diminishing as a global power, the United States stumbled through the turmoil of the 1970s-80s and kept on plugging. In 1991, President George H. W. Bush declared an end to the cold war. Whereas the Soviet military-industrial complex had collapsed under the pressure of trying to keep pace with its cold war nemesis (Rhodes, 2007), the United States joyfully declared victory in the cold war and, in the same breath, proclaimed itself the sole superpower in a new world order (Bose and Perotti, 2002). However, based upon the logic of industrial political-economics this made no sense. If the United States had declined as an industrial power, then its power in the global community should also be on the wane. However, by the 1990s, it was clear that something else was afoot.

Of course, what had changed was the fundamental nature of United States culture and economics. The ebbing of United States industrial production had, from the perspective of an industrial paradigm, seemed to indicate that America's power should also diminish. What such a perspective failed to account for was that, while America's industrial infrastructure was eroding, the United

States was laying the foundation for an entirely new socio-economic framework: the information society (Webster, 1995). This transformation was so huge and complicated that, for those in the midst of the transition, it was difficult to assess the merits of forswearing industrialism for a newly-emergent post-industrial society (Bell, 1976). Yet, if anything, developing into an information society amplified America's wealth, power and prestige.

Bill Clinton benefited throughout his presidency from the longest uninterrupted economic expansion in United States history. In part, the United States economy grew throughout the 1990s as a result of the post-cold war triumph of market economics. In addition, information technology came into its own as a defining feature of American culture during the 1990s (Okin, 2005). IT not only provided an entirely new type of fuel for the American economy, but the very nature of information technology seemed to create new opportunities for the United States to redefine its future. The cold war was old news. The problems associated with that struggle—that had seemed utterly irresolvable for the previous half century—had by the mid-1990s been reduced to little more than an afterthought. As always, old problems tend to look very different from the perspective of a new era. Certainly, serious concerns persist about how to manage the remaining cold war arsenals (Ritchie, 2009), but, for the most part, the United States and the rest of the world have moved on: there are new mountains to climb, new problematics to resolve.

Information has always been the key to solving problems. Knowing how to light fires has warmed a lot of fingers over the ages. Human super-adaptability is contingent upon the conscious application of information to the problem-solving process. In the effort to to develop super-adaptive solutions, information technologies have proven to be an exceptionally valuable new tool. Whereas cognition permits humans to rapidly invent super-adaptive solutions to environmental problems (i.e., humans can think more quickly than they can genetically adapt), IT substantially amplifies the quantity and quality of information that is available for problem-solving. Thus, with portable, wireless information technologies, it is possible to fill information gaps (e.g., finding a physician, translating foreign tongues, developing smarter information technologies, etc.) in a twinkling. Further, the prospect of developing AI could actually make the entire human problem-solving process antiquated.

Humans devote a great deal of time, energy and anxiety to solving problems. Arguably, should researchers ever be able to develop AI, it will no longer be necessary for humans to lose sleep over unresolved problems. The virtue of mechanization has always been that machines are stronger and more durable than their creators. In minutes, a steam shovel can do more work than a human can accomplish in a week. Presumably, the same will also be true for mechanized intelligence. Whereas, human arms and minds rapidly weary of the tasks to which they are assigned, machines do not. Thus, if or when researchers manage to develop AI, it will be possible to assign computers the daunting task of solving all of the wearisome problems (e.g., pandemic, climate change, food shortages, nuclear weapons management, energy crises, etc.) that humans have lacked the time, energy, or initiative to resolve for themselves. Indeed, the fact that machines could one day become more intellectually-gifted problem-solvers than humans has inspired Raymond Kurzweil to suggest that, if or when humans invent AI, then in that singular moment, the trajectory of life, humanity and intelligence will change forever (Kurzweil, 2005).

Kurzweil argues that, because of the mechanized work ethic described above, the moment that machines achieve a level of human-like sentience, those machines will almost certainly employ their newfound abilities to generate even higher orders of mechanized intelligence. As a result, although smart machines will be capable of solving a wide variety of extant problems, those same machines are likely to begin solving new problems of an entirely different magnitude. For starters, if humans ever invent machines that are smarter than their creators, then humans will in that same moment become only the second smartest beings on the planet. Given how poorly humans treat creatures of lesser intellect (e.g., we often serve them for supper), creating AI could end up being one of the dumbest ideas we've ever had.

This, by the way, is a recurrent theme in speculations about AI, i.e., if humans create machines that think for themselves, then what's to stop those machines from *thinking for themselves*? Since AI is a fantasy, it is somewhat irrational to fret unduly about the dangers that non-existent machines might one day pose. Yet, given the success that humans have enjoyed in resolving even the most formidable problematics, such an irrational anxiety may one day become a very real concern. Although there is no way to determine how grave a threat intelligent machines might one day pose to humans, such a prospect has done little to dissuade AI-related technological advancements. I suppose it's just one of those eventualities, like Hurricane Katrina, that we won't—and arguably can't—deal with until the sky starts falling.

Although Kurzweil is convinced that intelligent machines will one day outperform their human creators, he is not alarmed by that prospect. Kurzweil is convinced that the thought waves upon which human experience is based can be fully replicated in an artificial environment. Thus, just as one might make a duplicate copy of an MP3, Kurzweil believes that it will one day be possible to scan human brains and transfer their entire contents into the artificial cyberspace of smart machines (Kurzweil, 1999, p. 124). Rather than viewing such a transfer as somehow compromising humanity, Kurzweil believes that reanimating human intelligence in a machine environment might actually enhance the human experience. That is, smart computers will have the advantage of being intellectually more expansive, adaptable, collaborative and enduring. Thus, for Kurzweil, the prospect of reanimating human intelligence in a machine environment will change practically every aspect of what it means to be human, but the transformation, no matter how dramatic, will arguably constitute an improvement on the human condition.

I, for one, am not so certain.

Human-machine partnerships are never ideal, rather they are always problematic. I believe the same will be as true for increasingly intelligent machines as it has been for every other type of technology that humans have invented.

Ghosts in the Machine

Roboticized assembly lines changed industrial production profoundly. To a certain extent, roboticized factories represent the fulfillment of Frederick Taylor's (1911) vision of optimized industrial efficiency. For Taylor, efficiency involved breaking down complex tasks into their simplest constituent parts and then training factory laborers to work like automatons: to do simple jobs as efficiently as the machines with which workers were partnered. Industrial efficiency was great for the bottom line. More efficient factories helped workers generate a greater quantity of commodities. As a result, factory owners were able to secure more output for every dollar that they spent on labor.

Of course, not everyone agreed that Taylor's efficiency techniques were altogether beneficial. From a human perspective, Tayloristic efficiency is alienating. Not wishing to be treated like machines, factory laborers engaged in a wide variety of tactics, such as slow downs, sit downs and strikes, to emphasize that humans preferred to be treated better than machines (Venneti, 2003). With the help of the union movement, laborers and factory owners struck a working compromise: laborers agreed to work like machines, but only if their employers treated them like human beings.

This working compromise endured for much of the twentieth century. In the post-war era, United States corporations made staggering profits, while unionized laborers helped swell the ranks of the middle class. Of course, this relationship changed in the 1970s as the the seismic shift toward the post-industrial society got under way. Information technologies enabled conglomerates to coordinate international production networks and also made it possible to construct roboticized factories (Nof, 1999). Of course, one could argue that, if industrial production alienated laborers, then roboticized factories succeeded in liberating laborers from such drudgery. No longer did humans need to play the role of industrial automatons; automatons could play that role even better. The downside of this arrangement was that, in liberating humans from factory labor, the standard of living that industrial workers had struggled for the better part of a century to establish was severely undermined (Aamidor and Evanoff, 2010). Ever since the Luddite Movement (Bailey, 1998), it has been evident that, while generating gains for some, technological advancements often cause distress for others. Yet, time and again, the anguished protests of displaced laborers have fallen on deaf ears.

Generally speaking, progress involves a process of creative destruction (Schumpeter, 1975): the past must be demolished in order to blaze a path toward a brighter future, regardless of who gets hurt along the way. However, therein lies the rub. Must a better future necessarily be defined as a more technologically-intensive future? Even though it might be profitable to deploy new information technologies, I am not convinced that in every case, it is a good idea to do so. Take, for example, voice recognition technologies. Increasingly, voice recognition technologies have been deployed to replace human customer service representatives. Though capable of answering telephones and coordinating elementary information exchanges, I think it is safe to say that, at present, voice recognition technologies fall far short of serving as an adequate replacement for well-trained humans. Yet, regardless of the aggravation that half-baked voice recognition technologies may inflict upon millions of customers, these barely good enough technologies have become pervasive largely so that companies can squeeze a few extra bucks out of downsized customer service payrolls. The point that I would like to emphasize is that, instead of crafting information technologies that fulfill their true promise, IT developers have on more than once occasion created technologies that were just "good enough" (e.g., Microsoft Vista) and, in doing so, they have generated technological dystopias instead of utopias. Further, once having established sub-optimal operating standards, it can be exceedingly difficult to raise the bar.

Sub-Optimal Intelligence Standards

It is fair to say that craniometrists and psychometricians established sub-optimal standards for measuring intelligence. However, once having established those standards, it has been extremely difficult to alter the way that we define and measure intelligence. Alfred Binet is credited with being the originator of the intelligence test. Binet designed his test as a means to identify students who, being unable to succeed in typical classroom environments, might benefit from special educational opportunities. Binet went to some lengths to emphasize that his tests were not designed to evaluate the intellectual potential of cognitively well-functioning students. Utterly disregarding Binet's good advice, H. H. Goddard hawked Binet's tests in the United States as a means to establish the innate of intelligence of all test-takers (Gould, 1996, p.189). For Goddard, Binet's test consistently generated results that distinguished competent from incompetent test-takers and, as such, it served as a "good enough" measure of intelligence. To this day, standardized tests offer, at best, a woeful mismeasure of intelligence. Yet, in spite of their intrinsic deficiencies, the perception that standardized tests generate "good enough" measures of intellectual achievement remains entrenched, and this misperception has undermined the quality of education everywhere it has taken root (Meier, et al., 2004).

Getting back to the discussion of AI, Kurzweil has complained that many people fail to acknowledge that various types of pre-intelligent information technology in fact represent working versions of artificial intelligence (1992, p. 14). Frankly, I believe that the public's unwillingness to characterize extant technologies as manifestations of artificial intelligence is a good thing. As with the concept of intelligence, in any circumstance where we lower the bar of our expectations, the reality that we create tends to rise to the level or our expectations. Thus, if we begin referring to existing not-so-smart technologies as artificial intelligence, then progress towards an actual form of AI (i.e., technologies that are capable of passing the Turing test) will be derailed. In too many cases, half-baked smart technologies—such as the current generation of voice recognition software—have created more problems than they have solved: real human intelligence remains infinitely preferable to dumbed-down versions of AI.

The Aesthetics of Intelligence

Both Kurzweil (1992, p. 15) and Minsky (2006, p. 97) argue that the challenge of creating a full-fledged version of AI can be resolved by breaking complex issues into simpler problems: " . . . there is the faith in the AI community that most definable problems . . . can be solved, often by successively breaking them down into hierarchies of simpler problems" (Kurzweil, 1992, p. 15).

Certainly, the process of resolving problematics can often be advanced by breaking down large-scale problems into sequences of smaller, more manageable problems. For example, through the various phases of the Mercury, Gemini, and Apollo space programs, NASA executed a stepwise approach to sending astronauts to the moon. Each new achievement in space established a foundation upon which to aim even higher. Much the same has been true for the AI problematic. In 1950, Turing admitted that information technology would need to mature for half a century before computers would start to show glimmers of human-like intelligence. Following sixty years of innovation, computers have indeed come a long way: they are smaller, more powerful and more personal than most people could have imagined in 1950. Just as Turing predicted, computer speed, memory and software have advanced by leaps and bounds, but, as yet, none have come anywhere close to transcending the Turing-articulate threshold.

To this point, information technologies have advanced by breaking complex problems into sequences of smaller, more manageable problems; however, I do not think it will be possible to resolve the Turing problematic by "thinking smaller." In other words, as I have emphasized throughout this text, the most astounding new ideas generally do not emerge from solving intra-paradigmatic sequences of recognizable problems: building smaller and denser integrated circuits. Instead, new truths tend to emerge when clever people redefine reality. The precise source of creative inspiration, or agency, that makes it possible for individuals to redefine reality is unknown. As Minsky argues, brains don't make geniuses, it's what geniuses do with their brains that sets them apart (Minsky, 2006, p. 275). However, precisely how geniuses manage to achieve more gifted levels of brain function remains an abiding mystery. Nevertheless, such inspired, inventive thinking represents the essence of intelligence itself: intelligence is a mysterious, but wonderful quality of human brain function that equips humans with the ability to think and do things that no one has accomplished before.

Thus, I believe, to create artificial intelligence it will be necessary to invent entirely new ways to think about thinking. Intelligence is not so much a mechanical process as it is an aesthetic experience. Our intelligence derives from and informs our sense of identity: *I think, therefore I am*. Furthermore, Minsky and Kurzweil both assert that, as we think, humans have no experience of the mechanical, neurological processes that underlie cognition. In other words, humans rely upon their gray matter and the neurological activities that take place therein to engage in intelligent thinking. However, the act of thinking itself is not experienced as an infinity of neurological firings. Instead, humans, unless they are plagued by psychological disorders, typically experience cognition and intelligence as a seamless, singular experience. Intelligence is a cohesive compilation of an infinity of sensory inputs that combine to create a master identity that we recognize as our 'selves,' or our unique, sentient oneness.

That said, we need to know what intelligence is, and respect it in its broadest scope and potential, before we can hope to construct an artificial version that approximates human intelligence in a meaningful way. Thus, AI will need to incorporate the potential for reality redefining creative thought, which, at present, happens to be an extraordinary quality of cognition that humans are unable to fully grasp. Of course, the solution to that particular problem is to use human intelligence as an instrument to resolve the anomaly of human intelligence: we will have to redefine our conceptions of intelligence to develop a more adequate understanding of this mysterious cognitive phenomena. In other words, we can shed light upon what we do not presently understand about human intelligence by utilizing the enlightening qualities of human intelligence to modify the boundaries of the known universe in order to better understand the fundamental source of human intellectual creativity.

My feeling is that, if we are determined to create artificial intelligence, then we should do precisely that and nothing less. As we learned with ELIZA, it is certainly possible to create information technologies that masquerade as AI, but if we treat such chimeras as AI, then what have we really accomplished? AI will not exist until knowledge-seekers manage to resolve the Turing problematic. Technologies that fall short of the Turing threshold, while interesting and valuable in many ways, simply do not merit the honor of being called AI.

Intelligence is the most valuable resource that humans possess and it is a disservice to cheapen the concept in any way. Generations of craniometrists and psychometricians have devalued intelligence by inventing invalid measures of a phenomenon that defies simplistic measurement. Hopefully, AI researchers will not make the same mistake. If researchers are ever going to create a version of AI that is more than a mockery of human intelligence, then they will have to begin by grasping not merely the mechanics of intelligence, but its aesthetics. Intelligence is a sublime experience that is more than the sum of its parts. No machine that fails to grasp that essential fact will ever be able to fool a human interlocutor, nor should anyone presume to describe such a deficient mechanism as intelligent.

Haste Makes Waste

Kurzweil argues that intelligence is closely correlated with the speed of thought. He justifies this notion by asserting that psychometricians employ time as an evaluation criteria in their intelligence testing procedures (Kurzweil, 1992, p. 21; 1999, p. 44). In my view, nothing could be further from the truth. The more hurried our thinking, the less intelligent our intellectual output. For example, I have been teaching at universities for about twenty years, and on many occasions students have turned in papers which, by the students' own admission, have been a product of last minute preparation. In spite of having cranked out out a term paper in only a few short hours, students have often turned in their last-minute projects with the confident assurance that they (the students) "work well under pressure." However, in almost every case, the overall quality of last-minute output suffers dramatically in comparison to the scholarly efforts of students who have composed their papers more gradually. In emergency situations, such as when defusing a ticking time bomb, hurried thinking may be preferable to slower-paced thinking. However, in practically every other case, the more deliberate the thinking, the better the intellectual product.

Indeed, many of the greatest ideas have literally taken years to formulate. Although Einstein produced a spate of groundbreaking papers during his annus mirabilis, such a sudden outburst of intellectual productivity was predicated upon long years of cogitation. Einstein's genius was not the product of rapid thought, but of a fierce determination to ponder the deepest, most intractable mysteries of the universe until, bit by bit, he unraveled those enigmas. Thus, the best ideas are often those that take the longest time to formulate.

Just as human intelligence is not a product of computational speed, I think the same will be true of AI. Given the pace of IT development and the fierce determination with which the AI problematic has been attacked, I feel certain that AI developers will eventually create a type of machinery that resolves the Turing problematic. However, I do not believe that AI will spontaneously emerge from creating faster computers. No matter how many times microchip processing power doubles, intelligence will not emerge as a product of computing speed alone. It will only be possible to create artificial intelligence when humans finally figure out what it really means to be intelligent. To accomplish that landmark achievement, knowledge-seekers will have to think longer and harder about the true source of intelligence than they ever have before.

In closing, I challenge AI researchers to prove me wrong—or, better still, prove me right.

Chapter 6
Brightening a Dark Universe

For many, it might come as a surprise that Albert Einstein, a name that is synonymous with genius, formulated many of his greatest scientific breakthroughs while employed as a patent clerk (Isaacson, 2007). During his lifetime, Einstein became the world's most recognizable academic, but his fabled career got off to a rocky start. As an undergraduate, perhaps due to an unshakeable confidence in his own intellect, Einstein failed to display the type of deference that his professors expected from students. As a result, keen as he was to embark on an academic career, none of Einstein's professors were willing to support his bid for graduate studies. Being denied his foremost career preference, Einstein was forced to pursue other options (Ohanian, 2008).

Following an extended job search, Einstein landed a position as a patent clerk. So bleak had his employment prospects become that Einstein received news of the clerkship with great enthusiasm. Though the Swiss patent office was less illustrious than his academic aspirations, Einstein found patent work to be a pleasant diversion. Also, Einstein's modest workload left the greater part of his mental energies untaxed. As a result, even while fulfilling his duties as a patent clerk, Einstein could embark on extended contemplations of the most puzzling issues in theoretical physics.

So it was that Einstein labored in anonymity until 1905, Einstein's annus mirabilis, or his miracle year (Kaku, 2004). With the publication of four major papers during 1905, Einstein revolutionized scientific conceptions of the physical universe. In the first miracle year paper, Einstein argued that rather than flowing in unbroken streams, light energy was split into minuscule packets, called quanta. In the next paper, Einstein argued that the inexplicable Brownian motion of dust motes in fluids was due to the impact of particles that were so minute (i.e., atoms) few scholars had been willing to believe such infinitesimals could exist. In his third miracle year paper, Einstein introduced his theory of special relativity: a theory postulating that space and time are relative. And in the fourth paper, Einstein revealed the scientific equation that has come to immortalize his stunning theoretical achievements, $E=mc^2$. With this equation, Einstein established the fundamental equivalency of matter and energy and, what's more, he drew attention to the awesome and terrifying energy locked within atomic particles (Rhodes, 1986).

Though academic physicists might have been inclined to sneer, Einstein was fortunate that Max Planck was among the first readers of his revolutionary manuscripts. Max Planck, a renowned physicist and the Associate Editor of the *Annalen der Physik*, immediately recognized the genius in Einstein's theoretical inspirations. In part, because Einstein drew upon Planck's groundbreaking work in the realm of quantum physics, but also because of the sheer brilliance that Einstein exhibited in his articles, Planck dutifully published Einstein's submissions and, as a result, the physical universe has never looked the same since.

Prior to Einstein, Newtonian mechanics had dominated scientific conceptions of the physical universe (Hawking, 1988). In Newton's mechanical universe, time and space were fixed and believed to be measurable from an absolute perspective. In addition, until Einstein's miracle year, physicists had searched in vain for ether, the essential substrate of the cosmos. Ether was presumed to exist because, since light was understood to behave like a wave, Newtonian conceptions of matter and energy asserted that waves could only be propagated through a medium. Thus, for distant starlight to reach earth, space must be filled with ether. The problem was, try as they might, scientists were unable to locate the existence of ether. Though unthinkable, it almost seemed as if in the deep, dark void of space *there was no ether*.

In Einstein's universe, ether was not only non-existent, but it was unnecessary. Einstein dispensed with this key feature of the Newtonian universe by asserting that light exhibited properties of both waves and particles. In addition, Einstein repudiated Newton's conviction that time and space were fixed by envisioning an entirely new convergence of matter and energy. Einstein argued that, though the speed of light was always constant, space and time could be warped depending upon one's relative velocity. Consequently, all that had once been certain in Newton's mechanical universe had suddenly become relative. Einstein had literally redefined the fabric of reality.

Just as many great thinkers had done before, Einstein encountered an inescapable dissonance between his observations of the empirical universe and established scientific theories—in this case, Newtonian mechanics. Faced with this dilemma, Einstein drew upon his exceptional creativity to construct new theories that would render previously inscrutable phenomena intelligible. So, like other great thinkers, Einstein stretched the theoretical boundaries of his discipline in order to bring scientific knowledge into closer accord with the parameters of the known universe. In so doing, one could argue that Einstein developed a more truthful lens through which to view the physical universe.

On the other hand, in sweeping aside Newton's mechanical universe, one could also argue that Einstein illustrated that science is a fundamentally relativistic endeavor (Ladyman, 2002). In other words, for over two hundred years, Newton's mechanical paradigm had represented The Truth about the physical universe. Yet, in the short span of his Miracle Year, Einstein undermined many of Newton's key presuppositions. After 1905, there was a new standard for determining truth in the cosmos. Newton was old news. Essential as such a paradigm shift may have been, nevertheless, the transition from Newton's to Einstein's ideas cast a troubling light on the role of truth in science. Essentially, one had to ask, was there anything especially truthful about scientific truths? In other words, if universal truths are only valid relative to particular scientific paradigms and, in turn, if paradigms are fleeting and disposable, then do scientific truths represent anything more than intellectual fads?

Perhaps the simplest definition of truth is "knowledge that unbiased, undistorted, or pristine" (McGettigan, 2006). Thus, truth can be understood as knowledge that describes an unvarying quality of the empirical universe. For example, Einstein's 1905 discourse on the photoelectric effect asserts that the most elemental unit of luminous energy is a particle-like quantum known as a photon. The importance of this or any other statement of truth is that, first of all, the statement reveals something useful, intelligible and demonstrable about the universe: light travels in, and can be measured in photons. Another crucial feature of scientific truth is that it must represent knowledge that remains valid across time and space. Whether on earth, Mars, or in the Andromeda galaxy, light must travel in quantum-sized photons. Therefore, regardless of the time or place, truth must be universal and unvarying.

Getting back to the transition from Newtonian mechanics to Einstein's relativity, one must wonder: if science is, first and foremost, an endeavor concerned with the pursuit of truth, then how can scientific paradigms be disposable? That is, many of the truths touted by Newtonian mechanics (e.g., time and space are fixed) are, according to the relativity paradigm, demonstrably false. Consequently, if the knowledge contained in any particular scientific paradigm is demonstrably false, then what guarantee is there that the truths touted by any paradigm, including Einstein's relativity perspective, are any more reliable? Indeed, critics of science have argued that, given the limitations of human cognition and the imperfections in the knowledge systems that we create, it is simply impossible for science to capture universal, unvarying truths (Denzin and Lincoln, 1994; Foucault, 1980). No matter how earnest scientists may be in their pursuit of truth, the limitations on the human ability to conceive and represent Ultimate Truth makes it impossible to generate knowledge that approximates what is really, finally, universally true. Indeed, critics of science have pointed to the common occurrence of paradigm shifts as an illustration of this problem: the most fundamental truths endorsed by one paradigm (e.g., time and space are fixed, or the void of space is filled with ether, or the earth is the center of the universe, etc.) are often viewed as examples of rustic ignorance by succeeding paradigms. As such, critics of science have argued that far from representing timeless, transcendent knowledge, any truths that science may proclaim are strictly relative to, and delimited by, the dominant paradigms of the day (Lemert, 1991).

It is indisputable that the scientific truths of one era often tend to be the butt of jokes in the next. From the perspective of the 21st century, it seems laughable that anyone could ever have believed that the earth was flat (Garwood, 2008), or that it lay at the center of the universe. Yet, in spite of the relative impermanency of scientific truths, I argue that it is possible to identify at least a grain of truth in even in the most antiquated scientific paradigms. For example, the ancient Greeks believed that the universe was constituted of four basic elements: earth, air, fire and water (Bakalis, 2005). Of course, contemporary periodic tables illustrate that there are far more elemental substances than the ancient Greeks could ever have imagined. Nevertheless, were the Greeks completely deluded to think that there were only four basic elements? In fact, I would argue that the ancient Greek paradigm was relatively truthful in that each of the four elements manifested demonstrable properties of "creation," e.g., a mixture of water and earth brought forth life in farmers' fields, volcanic fire routinely created new land masses, and air was indisputably essential to sustain life. Certainly, in the 21st century, we know a great deal more about the elemental forces that shape the universe, but I think it is still fair to say that the Greeks were not delusional when they asserted that the universe was constituted of four basic elements. Rather, they were merely operating within a paradigm that was designed to make sense of the available evidence at the time.

That scientific paradigms are imperfect is not a matter of dispute. No matter how cherished, no matter how seemingly flawless, no matter how groundbreaking, the simple fact is that every scientific paradigm that humans have ever invented—or, I would hazard, that humans ever will invent—is flawed. This is the case, quite simply, because no matter how exacting scientific research may be, the human capacity to observe and conceptualize truths is limited. The universe is boundless and full of surprises, whereas the minds that seek to compass that selfsame infinitude are limited and flawed. Thus, scientific paradigms can never incorporate the "entire truth." There will always be more to know. If scientists do their work properly, scientific paradigms will always be short lived.

That said, I disagree with those who argue that science is an utterly relativistic endeavor. In contrast with Einstein's relativity paradigm, relativism is a perspective which asserts that all knowledge is equally (in)valid. Relativists assert that there is no such thing as truth (Seidman, 1991; Richardson, 1998; Clough, 1994). Such an assertion is patently false. First of all, it must be understood that there is a distinction between truth and its subsequent perception and representation by humans. Though solipsists would insist otherwise, it is important to acknowledge that there is a reality "out there" (Lyng and Franks, 2002). Just try to deny it the next time you bump your head on a tree branch. In brief, Karl Popper has noted that such intellectual discord is rooted in the fundamental dissonance between realist and idealist philosophy. Popper (1982b) asserts that realist philosophers accept that empirical reality exists independently of the perceptions of individual observers. In other words, realists believe that there is a real world *out there*. By contrast, idealists argue that reality is only accessible via subjective perceptions. As a result, it is impossible to demonstrate that anything real, substantive or empirical exists outside an observer's subjective perceptions. In the most extreme version of such an idealist perspective, the universe is presumed to be nothing more than a figment of the imagination of individual observers (Clegg, 2009). Interestingly, Einstein's negative reaction to such a perspective in the field of physics led to one of the greatest scientific debates in the twentieth century.

Quantum Entanglement

By building upon Max Planck's discoveries, Einstein pointed the way toward an entirely new intellectual frontier in the realm of sub-atomic physics. While Einstein devoted the bulk of his attention to developing his theory of general relativity, a generation of trailblazing young physicists surged into the bizarre landscape of quantum physics. Chief among this new breed of physicists was Niels Bohr, who established an institute at the University of Copenhagen that was entirely devoted to the advancement of quantum physics. As Bohr and his institute attracted many of the greatest physicists of his age, their collaborations ultimately produced what has come to be known as the Copenhagen interpretation of quantum mechanics. Once established, the Copenhagen interpretation became the dominant paradigm in the field of quantum physics, but as the details of this new theoretical perspective came to light they were greeted with deep skepticism by a towering figure in physics, none other than Albert Einstein.

Essentially, the Copenhagen interpretation endeavored to account for the puzzling duality, or what Bohr would later identify as the complementarity, of phenomena at the sub-atomic level. Again, Einstein was among the first to recognize such elemental dualities. For example, though it seemed to defy reason, Einstein postulated that light exhibited qualities of being both a wave and a particle. Up to the time of Einstein's astounding discovery, classical physicists were convinced that physical phenomena had well-defined, distinct and consistent properties. Thus, there had previously been a clear separation between matter and energy, electricity and magnetism, time and space, particles and waves, etc. However, many of Einstein's breakthroughs revealed that each of these phenomena, distinct as they may have seemed, were nevertheless, integrally related. Willing as Einstein may have been to acknowledge such mind-bending dualisms in his own work, he was nonetheless discomfited by the conclusions drawn by Bohr and his collaborators about quantum phenomena. Early on, the most objectionable elements of the Copenhagen interpretation for Einstein derived from Heisenberg's (1930) uncertainty principle.

Heisenberg argued that, from the perspective of classical physics (meaning physics before Einstein), every object in the universe was presumed to have a precise mass and momentum. These facts could easily be established through observation: mass and momentum could be specified by using straightforward measurement procedures. Furthermore, classical physics asserted that objects, such as billiard balls, would maintain their mass and momentum regardless of whether they were subject to direct observation or not. For example, if someone were to fire a cannon, though observers might not be able to examine the projectile in flight, it would still be possible to infer the cannon ball's trajectory by taking issues such as mass, acceleration and gravity into account. The final proof of the projectile's flight could be established by traveling to the impact site and recovering the cannon ball. Although Heisenberg was willing to concede that the above statements were generally true for conventionally observable phenomena, Heisenberg asserted that everything changed at the level of the quantum.

To begin with, quantum phenomena are not observable in the conventional sense. Although we can observe billiard balls with our eyes, quantum particles are so minute that they defy optical observation. Thus, the only way to observe quantum particles is to employ measurement techniques that interfere with the particles. As a result, Heisenberg argued that precise measures of a quantum particle's position tend to interfere with a particle's momentum, whereas measures of momentum tend to modify position. Though the idea initially made him squeamish, Heisenberg (1930) concluded that the exactitude of classical physics must yield to uncertainty in the quantum realm: it is impossible to determine both the precise momentum and position of quantum particles. In turn, as Bohr expanded upon Heisenberg's uncertainty principle, the Copenhagen interpretation opened a chasmic philosophical divide in the physics community.

At the 1927 Solvay Conference, Bohr unveiled his full-blown theory of quantum mechanics. Bohr introduced this radical new perspective by making the relatively banal assertion that the only way to establish a quantum object's characteristics (e.g., position vs. momentum, or wave vs. particle) was to make observations. However, according to Bohr's (1928) quantum postulate, in every case, the act of observation modified the dualistic characteristics of the quantum phenomenon in question. Put somewhat differently, Bohr argued that there was no clear separation between observers and the phenomena that they observed. As such, Bohr added that there was no longer any way to assert the unequivocal existence of objective reality, "an independent reality in the ordinary physical sense can neither be ascribed to the phenomenon nor to the agencies of observation" (Bohr, 1928, p. 54). Thus, Bohr asserted that studies which were designed to search for the wave-like features of quantum particles generally produced results which indicated that quantum particles were wave-like. The same was true for studies designed to focus on the particle-like qualities of quantum phenomena. As a result, Bohr reformulated Heisenberg's uncertainty principle to make the more outlandish assertion that it was the observation process itself that modified quantum phenomena. It was almost as if quantum particles somehow "knew" that observers were intent upon witnessing a particular type of quantum activity and then modified their behavior accordingly. As if that weren't enough, Bohr still had one more nail to hammer into the coffin of realism. Bohr's coup de grace was to argue that, because quantum states are dependent upon and vary in relation to observation, if a particle is not subject to observation then one should not assume that it has either position or momentum. Even more, Bohr's newly codified Copenhagen interpretation adopted the anti-classical position that, "An unobserved electron does not exist" (Kumar, 2008, p. 262).

For Einstein, Bohr's final lurch into solipsism was too much to bear. Einstein was convinced that the physical universe was more objective, solid and precise than the Copenhagen interpretation proposed. Further, Einstein believed that the singular purpose of science was to unlock the secrets of the physical universe. Years later, Einstein affirmed this conviction in a letter to Max Born, "You believe in a God who plays dice, and I in complete law and order in a world which objectively exists" (Kumar, 2008, p. 331). In short, Einstein was a realist and he insisted that an objective universe must exist independently of the cognition of individual observers. For example, even though Einstein had not witnessed the eclipse that Arthur Eddington used to test his theory of relativity, Einstein was convinced that the eclipse had actually taken place: the eclipse was a real phenomenon in an objective universe that was governed by universal physical laws. From Einstein's perspective, even though quantum particles might be substantially smaller than the earth, sun, and moon, nevertheless, quantum particles were every bit as real as larger physical objects. Therefore, if the earth, sun and moon had identifiable properties of position and momentum, then the same also had to be true of quantum phenomena. Consequently, Einstein was convinced that, if Bohr and his collaborators were not able to determine both the position and momentum of quantum particles, then quantum mechanics could not be a complete theory. Further, when objectively-minded physicists decided to develop a complete theory of quantum mechanics, Einstein insisted, all that was fuzzy in the empirical universe would become crystal clear.

For their part, Bohr, Heisenberg and other supporters of the Copenhagen interpretation argued that quantum mechanics was a complete theory. In other words, due to the essential dualisms that were intrinsic to quantum reality, supporters of the Copenhagen interpretation argued that quantum mechanics offered as final and thorough a statement on the quantum realm as it was possible to articulate. Unsatisfied with that response, Einstein drew upon his vast intellect to design one fiendishly clever thought experiment after another. Each of the thought experiments was intended to emphasize the incompleteness of the Copenhagen interpretation. Einstein was convinced that a more complete theory would surely re-introduce sanity into the conceptual morass that was quantum physics.

In spite of Einstein's insistence upon a realistic interpretation of quantum mechanics, his primary antagonist, Neils Bohr, successfully deflected each of Einstein's conceptual assaults on the Copenhagen interpretation. Strangely, the more insistently that Einstein defended his realistic principles, the more Einstein came to be viewed as a stodgy old has-been. After settling in at Princeton University, Einstein confided to Max Born that, among his new colleagues, he was "considered an old fool" (Born, 2005, p. 128).

The Thought Experiment from Hell

Still, Einstein stuck to his guns. Making one last titanic effort, and working with two collaborators, Einstein designed a thought experiment that even Bohr was never able to fully debunk (Einstein, Podolsky and Rosen, 1935). As noted above, Einstein objected to the quantum-mechanical notion that, in the absence of measurement, particles had neither position nor momentum. Thus, in his EPR collaboration, Einstein set out to demonstrate that it was possible to determine, whether subject to observation or not, particles possessed real physical properties, such as position and momentum. Importantly, the EPR argument accepted the quantum-mechanical contention that it was not possible to specify position and momentum simultaneously. Instead, the EPR argument simply endeavored to affirm that position and momentum were real properties of real particles in an objective reality. If the EPR paper could establish that particles were imbued with real, identifiable properties for which quantum mechanics could not account, then Einstein would finally be able to prove that quantum mechanics was an incomplete theory.

Thus, the EPR thought experiment involved two particles, or systems that interacted for a period of time and then flew off in different directions. Once the particles were separated, it would arguably be possible to obtain precise measures of particle I without interfering with particle II. Einstein insisted that this was the case because of a phenomenon that he referred to as locality. In other words, Einstein believed that for particles to interact, they needed to be in the same vicinity. According to Einstein's concept of locality, once particles have separated, they no longer have any direct influence upon each other: after separation, modifications to the position or momentum of particle I would have no influence on particle II. To suggest that separated particles might continue to interact would imply that particles were somehow endowed with the "spooky" ability to communicate instantaneously over great distances. Not only would such bizarre interaction defy common sense, but it would also require faster than light communication, which would violate Einstein's theory of relativity. Ergo, EPR concluded that instantaneous action-at-a-distance was not a reasonable feature of reality.

Proceeding from Einstein's assumptions about locality, after particles I and II have become separated, any measurement of particle I would presumably have zero impact on particle II. Therefore, EPR asserted that, equipped with knowledge about the original particle interaction, it would be possible to measure particle I's position and then make precise inferences regarding particle II's position. The same would also be true for measures of momentum. Certainly, EPR conceded, it would not be possible to make precise simultaneous measurements of particle I's position *and* momentum. Nevertheless, EPR asserted that, because quantum mechanics asserted that it was possible to establish either the precise position or momentum of particle I, then it must also be possible to establish that particle II possessed those same attributes. As a result, EPR drew the conclusion that, even though particle II was unobserved (and therefore undisturbed), its position, momentum and its very existence were confirmed elements of objective reality. Since quantum mechanics denied that unobserved particles existed in an objective reality, and therefore could not account for such phenomena, EPR concluded that quantum mechanics was an incomplete theory.

Upon publication of the EPR paper, Einstein once again succeeded in sending shock waves throughout the physics community. Although Einstein was perceived by many physicists as a tiresome curmudgeon (Folsing, 1997, p. 699), Bohr approached Einstein's EPR critique as a mind-bending but essential test of quantum mechanics. In the end, Bohr replied to Einstein in much the same way that Hamlet addressed his naive compadre, Horatio: "There are more things in heaven and earth, Horatio, Than are dreamt of in your philosophy." While Bohr was willing to concede that spatially-distant particles might not be able to disturb each other mechanically, nevertheless, Bohr insisted that the process of observing the position or momentum of one particle had what he referred to as an "*influence on the very conditions which define the possible types of predictions regarding the further behavior of the system*" (Bohr, 1935, p. 700, italics in original). That is, Bohr believed that particles, such as those Einstein described in his EPR thought experiment, that interacted at one point in time would remain entangled in a single system, and, thus, would exert a mysterious but nonetheless instantaneous influence on each other even when the particles became distantly separated. Again, Bohr linked this type of quantum entanglement to observation; arguing, that humans modify reality—often in the most puzzling and unexpected ways—in the process of observing reality. Thus, Bohr rejected the EPR argument on the basis of his quantum postulate: for humans, reality does not exist until we find a way to observe it, and in the process of observing reality, humans invariably modify it.

To say the least, Einstein was not impressed. He and Bohr were clearly at an impasse. More than once, Einstein spoke disparagingly of Bohr's suggestion that separated particles could somehow influence each other, referring to such quantum entanglement as "spooky action at a distance" (Born, 2005). For Einstein, the universe was neither the product of individual imaginations, nor was it subject to change based upon the vagaries of human perception. In short, God did not play dice with Einstein's universe. Einstein was convinced that a right-minded God would only create a universe that was logically consistent from the behavior of the smallest sub-atomic particles to the largest supra-galactic phenomena. What's more, God's version of the universe should ring true with Einstein's expectations of reality, or else, Einstein was convinced, *there was something wrong with reality!*

For his part, Bohr was swayed by the troubling, but nonetheless, compelling complementarity of quantum phenomena. There seemed to be no way around it: at the quantum level, the universe was a wacky place. Interestingly, as much as Einstein insisted that his was a more reasonable, and scientifically orthodox definition of the universe, the facts (strange as they were) tended to affirm Bohr's version of quantum mechanics. Again and again, experimental results have supported the Copenhagen interpretation of quantum mechanics. Yet, although

Einstein grudgingly acknowledged that quantum mechanics was the best available theory to explain sub-atomic physical phenomena (Einstein, 1950), he remained insistent that it was an incomplete theory. Einstein and Bohr remained conceptually and collegially at loggerheads until their final days: Einstein whiled away his final years contemplating an elusive unified field theory, while Bohr literally drew his last breath while wrestling with one of Einstein's more nettlesome thought experiments (Kumar, 2008, p. 327).

Complementary Realities

So, the question remains: Which of these two titans of modern science was correct? Certainly, their ideas regarding quantum reality were contradictory and mutually exclusive. Consequently, in a rational universe, one would expect that one of the two, either Bohr or Einstein, would have to be correct, while the other was demonstrably wrong. However, in the finest tradition of quantum complementarity, I believe that, precisely because of their irreconcilable differences, Bohr and Einstein were each both right and wrong.

Just as quantum particles exhibit dualistic and yet complementary characteristics, I believe the same is also true for the theories that describe quantum reality. Though Einstein and his supporters may have been loath to admit it, Bohr's version of quantum mechanics has been supported again and again by scientific observation. Irrational as it may seem, quantum particles consistently exhibit spooky action at a distance (Clegg, 2006; Kumar, 2008, p. 350). Precisely why such a phenomenon should exist remains a mystery, but the accumulated evidence of repeated observations is difficult to deny. The act of observation seems to influence the phenomena that are the subject of observation. While such an idea has troubling implications for those who believe in an independent, objective reality, Bohr's quantum postulate does expose a fundamental truism about science. Knowledge and facts are intellectual constructs: we know only those things that our minds are capable of perceiving. What humans experience as reality is essentially a figment of the imagination. Thus, reality may exist out there, but, then again, it may not. However unlikely, there is no way to prove that reality is anything more than that which the mind perceives. That being the case, it stands to reason that, at least on some level, phenomena must always be influenced by the processes of observation. This is as true for quantum particles as it is for human social behavior (Landsberger, 1958). Observation influences reality.

True as that may be, it is still easy to appreciate Einstein's antipathy for quantum mechanics. The idea that, on any level, objective reality might not exist independently of observation, or that reality might be subject to modification at the whim of observers is dissatisfying. Can observers truly exercise so much power over the constitution of reality? The idea seems to defy common sense. Rather than transforming to comply with observers' hopes and expectations, reality tends to yield only grudgingly in the face of individual initiative. For example, try as I might, I have never been able to melt the frost off of my windshield by dreaming of balmier weather. Thus, there is abundant evidence to support faith in an observer-independent reality: just because I take my eyes off of the sun does not mean that it will cease to exist. Whether our name happens to be Horatio or not, humans regularly encounter new phenomena that do not fit within our current philosophy. Thus, as Einstein argued, it is the task of science to explain the endless peculiarities of an objective universe over which humans have only limited understanding and control. Science is a knowledge-seeking enterprise that is forever in a process of reinventing itself. Thus, Einstein was actually less troubled by the bizarre behavior of quantum particles than he was by Bohr and Heisenberg's suggestion that quantum mechanics was a complete theory.

To suggest that quantum mechanics is a complete theory is to conclude that science has arrived at the final, ultimate truth about quantum reality. Again, due to the uncertainty associated with quantum phenomena, Bohr and Heisenberg believed that it would never be possible to develop more precise knowledge about the quantum realm than the Copenhagen interpretation would allow. Einstein objected to this conclusion largely because the theory of quantum mechanics itself acknowledges that it can provide at best an imperfect description of quantum phenomena. Thus, Einstein was willing to concede that quantum mechanics might be the best available theory to describe quantum phenomena, but he felt certain that, due to its inherent limitations, quantum mechanics offered by no means the final word on quantum reality. If quantum mechanics offered an imperfect description of quantum reality, then Einstein believed that scientists bore the responsibility of developing better theories. Though he became an object of ridicule in the physics community—indeed, Clauser reported that colleagues regularly assured him that Einstein's objections to quantum mechanics were evidence that he had gone senile (Bertlmann and Zeilinger, 2002, p. 72)—Einstein remained adamant that scientific journeys did not end with mysteries, rather, he insisted that's where they should begin.

Although I am not a physicist, I support Einstein's principled objection to the idea that quantum mechanics is a complete theory. However, my objection does not derive from any specific objection to quantum mechanics, rather I object on the grounds that any theory that has been constructed by imperfect intellects must, by its very nature, be imperfect—or, in the language of the EPR debate, incomplete. By this, I do not mean to suggest that science is an utterly futile endeavor, i.e., since humans cannot generate perfect or truthful knowledge, then there is no point in even attempting to do so. Far from it, I agree with Karl Popper who argued that knowledge acquisition is a never-ending process. Whatever we know now, regardless of the field of endeavor, is less than we can eventually learn and, therefore, represents less than we ought to know. As science advances, humans benefit in myriad and often unexpected ways. Science is at its best when its practitioners assume that *all* existing knowledge is incomplete. Thus, by doubting the completeness of quantum mechanics, Einstein was neither deluded nor senile, he was simply being a good scientist. Nonetheless, all too often, knowledge-seekers who challenge beliefs that other scientists hold sacred tend to encounter violent opposition rather than collegial encouragement.

Any truth-seeking endeavor will cease and fail at the very moment that knowledge-seekers decide that they have arrived at final, ultimate truths. This is the case because, first of all, it is simply impossible for imperfect minds to compass universal, ultimate truths. This point is particularly well-illustrated by the fact that earthlings currently enjoy only an imperfect understanding of a minuscule fraction of the universe (Clegg, 2009; Overbye, 2006). Perhaps even more problematic: those who assume that they have captured final, ultimate truths often intentionally close their minds to further inquiry (Popper, 1957). Thus, not only do true-believers invest faith in knowledge that represents anything but the Ultimate Truth, but they also refuse to entertain ideas that might challenge their preferred assumptions. Nothing could be more antithetical to the goals of good science.

Like all creatures, for humans to survive, they must adapt. However, human adaptation is tied to pursuing problematics that tend to modify the very nature of truth and reality. Bohr was right: observers redefine reality—and often in ways that they might neither choose, nor prefer. As Einstein discovered, new truths can often be traumatizing; however, the greatest scientists pursue the truth no matter where it leads, and no matter what the implications might be for their cherished beliefs or scholarly reputations. As Popper argued so persuasively, solutions are naught but the gateways to new problems. Those who are convinced that existing knowledge is far from perfect will always be the wisest of us all.

In the never-ending search for truth, I prefer the company of heretics.

Chapter 7
The Future is Problematic

"The significant problems we face cannot be solved at the same level of thinking we were at when we created them"—Albert Einstein

As we assess the current state of global affairs, it is easy to become disheartened. Problems that have plagued human society—poverty, hunger, disease, conflict, etc.—seem to grow more serious with each passing year. Yet, humans have always confronted seemingly insoluble problems. As Karl Popper argues, solving problems intellectually is a uniquely human endeavor. Most life forms solve problems with their biology, whereas humans solve problems with their brains. And it is through an iterative problem-solving process—which Popper describes as a process of conjecture and refutation—that humans have been able to super-adapt to an unprecedented variety of survival challenges. Yet, as Albert Einstein emphasized, it must be understood that any solution to a particular problem simply creates new problems. For example, the conveniences associated with life in the modern world have facilitated exponential population increases. In turn, exploding populations have introduced titanic stresses on the biosphere.

Further, Einstein points out that the problems humans create tend to exceed "the level of thinking we were at when we created them." Frustratingly, this means the reward for solving problems is often the invention of even more intractable problems. For example, Oppenheimer and the other Manhattan Project scientists solved the problem of how to win WWII by inventing weapons powerful enough to extinguish all life as we know it. As such, Einstein astutely identified a basic dissonance between problem-solvers and the problems that they invariably create. It takes genius to split atoms, but it takes an even higher order of genius to split atoms safely.

A Glass Half Empty?

Certainly, one could interpret Einstein's comments pessimistically, i.e., humans will never be equal to the problems that they create. The most imaginative solutions to intractable problems will always generate difficulties of even greater magnitude, e.g., solving the 19th century problem of whale oil shortages by tapping into the earth's vast petroleum reserves. It's as if the seeds of our destruction are built into every stroke of genius that's designed to save our bacon. The better the idea, the more disastrous the unforeseen outcomes. From this perspective, not only is the glass half full, but we might be better off spilling its contents on the ground. Luckily, it's also possible to interpret Einstein's admonition optimistically.

Humans have a penchant for creating problems. By their very nature, problems can be understood as challenges for which there are no established solutions. For example, at present, there is no way to meet global energy needs without increasing dependence on fossil fuels. Thus, pessimists would tend to view problems as irresolvable obstructions: if humans lack the necessary intellectual resources to solve problems, then we're stuck. It's simply not reasonable to expect people to fix problems that require more brain power to solve than humans can currently muster—or is it?

Our Cup Runneth Over

How can humans avoid destroying themselves if, every time we solve a problem, we accidentally create new problems of even greater magnitude? The answer: Humans need to get smarter.

Certainly, it's a well-established fact that humans can elevate their thinking. We do it all the time. Thus, the Wright Brothers' mastery of the skies, and NASA's conquest of the moon. However, Einstein's observation points to an even more fundamental dynamic. Everyone must get used to the idea that there will never be final solutions in any field of human endeavor. No matter what problems we may solve, the one thing we can be sure of is that even more mind-bending problems will rapidly ensue. At a glance, this assertion may seem to head in much the same direction as the pessimistic perspective described above: the day will inevitably arrive when humans are overwhelmed by insuperable problems that, like *The Terminator*, are, tragically and ironically, the unintended outcomes of their own brain children.

The long parade of great civilizations that have risen and fallen illustrates that this is not an idle anxiety. Whether the culprit is food, disease, war or whatever, societies that fail to pursue proactive solutions to their most pressing problems do so at their peril. The jury may still be out on precisely what precipitated the demise of the Roman Empire, however, what's certain is that, in the end, the Roman's were simply no match for the problems that their empire created.

Gloomy as the above realities may be, problems should not be perceived solely as harbingers of doom. As I have argued throughout this book, problems often serve as agency-motivators. Again, human agency emanates from a fundamental wellspring of intellectual creativity and equips people with the ability to redefine reality. Redefining reality is an intellectual process whereby people proactively develop solutions for unresolved problems, e.g., finding suitable alternatives to fossil fuels, or creating machines that think. Thus, redefining reality not only helps agents create proactive solutions to challenging problems, but it also expands and improves human intellectual horizons.

Reaching for the Stars

Knowledge-seekers begin the problem-solving process at a point where, as Einstein pointed out, they are not quite equal to the challenge at hand. However, as the problem-solving process unfolds, the status of the problem-solver viz. the problem du jour undergoes a transformation. In addition to generating a solution to a challenging intellectual problem (e.g., arguably, humans have benefited from the basic insight that round wheels help wagons roll along), the problem-solving process also elevates the problem-solver's intellectual capabilities, i.e., Kennedy's moon-landing problematic stimulated the United States to reinvent itself as the leader of the information society.

As a result, one can argue that solving problems helps humans accomplish more and better things than they ever could otherwise. That being the case, the fact that humans are not initially equal to the problems that they create can be viewed, not as a slur upon humanity, but as an invaluable opportunity for humans to *progress*. Being unequal to problems often stimulates naturally curious knowledge-seekers to master those problems. Indeed, even in situations where the practical benefits of solving a problem are uncertain, knowledge-seekers have often forged ahead simply because they despise the idea of being bested by a challenge. This was certainly the case with Einstein, who labored valiantly until the end of his days to formulate an elusive unified field theory.

Thus, rather than being humanity's undoing, problems promise to pave the way to a better and brighter future, but only if knowledge-seekers recognize that they are the masters of their own fate and, thus, must remain eternally committed to the challenge of solving the next set of intractable problems. In the problem-solving endeavor, there is no time to rest on one's laurels. Just because the United States managed to blaze a path to the information society does not mean that America's competitors will pass up an opportunity to usurp US leadership. As Einstein pointed out, past success only ensures that one must grapple with even more challenging problems in the future.

Take, for example, the Malthusian dilemma. In the eighteenth century, Thomas Malthus predicted the imminent demise of humanity. Malthus was convinced that the geometric rate of population growth would precipitate a catastrophic population crash. Malthus was correct in that, using eighteenth century technology, population growth would soon have been unsustainable. Of course, Malthus failed to account for the transformations that would soon accompany the industrial revolution. Had humans applied pre-industrial thinking to the problem of population growth, in all likelihood, the Malthusian disaster would have transpired. However, due to the cultural and scientific advancements associated with the industrial revolution, humanity has thus far avoided the fate that Malthus predicted.

As with many prognosticators, Malthus failed to see the future coming. That is, Malthus viewed the future through a paradigm that was better suited to make sense of the past. When Malthus assumed that the future would inevitably be shaped by the same forces that had defined his present, he made a critical error. The problems that plague one historical era, so long as humans succeed in elevating their thinking, often tend to be viewed as relatively minor challenges in succeeding eras.

This is not to say that the advancement of scientific knowledge will gradually create a perfect world. Far from it. In agreement with Popper, I believe that the problem-solving process is never-ending. Each solution will always introduce an entirely new set of even more challenging problems. The advantage of the unending quest for scientific truth is that it elevates and inspires new thinking. New problems not only create opportunities to seek new truths, but they also transform the substance of reality. As JFK discovered, there is nothing like a seemingly impossible goal to bring out the best in everyone.

Ludwig Wittgenstein once stated, "What we cannot speak about we must pass over in silence" (Wittgenstein, 1961, p. 62). What a bunch of nonsense. Although I agree that there will always be intellectual challenges that exceed the *existing* capacity of human intellectual inquiry—ideas for which we currently lack the ability to think and discuss intelligibly—those unknowns merely represent a new threshold of problems that humans must endeavor to resolve. If we were to adopt Wittgenstein's defeatist attitude (i.e., if we can't currently grasp an idea, then we shouldn't even try), humans would never take up new challenges. That is simply unacceptable. Impossible goals tend to be the ones that are most worth shooting for.

Therefore, the most certain route to a brighter future is to ignore the advice of defeatists like Wittgenstein and set our sights on the most remote star in the universe. We may never get there, but we'll sure as heck benefit from the journey.

Bibliography

Aamidor, Abe, and Ted Evanoff. *At the Crossroads: Middle America and the Battle to Save the Car Industry*. Toronto, Ontario, Canada: ECW Press, 2010.

Ackerman, Paul D. *It's a Young World After All: Exciting Evidences for Recent Creation*. Grand Rapids, MI: Baker Books, 1986.

Agassiz, Louis. *A Journey in Brazil*. Boston: Ticknor and Fields, 1868.

Agassiz, Louis, and Elizabeth Cary Agassiz. *Louis Agassiz: His Life and Correspondence*. New York: Houghton, Mifflin and Company, 1885.

Althusser, L. "Contradiction and Overdetermination," in *For Marx*, translated by Ben Brewster, pp. 87-128. New York: Verso, 1969.

Angelo, Jr., Joseph A. *Human Spaceflight*. New York: Facts on File, Inc., 2007.

Ashton, Thomas S. *The Industrial Revolution (1760-1830)*. Oxford: Oxford University Press, 1948.

Bailey, Brian J. *The Luddite Rebellion*. New York: New York University Press, 1998.

Bakalis, Nikolaos. *Handbook of Greek Philosophy: From Thales to the Stoics*. Victoria, BC: Trafford, 2005.

Barnett, Vincent. "Understanding Stalinism: The 'Orwellian Discrepancy' and the 'Rational Choice Dictator'." *Europe-Asia Studies* (May, 2006).

Barrow, Mark V. *Nature's Ghosts: Confronting Extinction from the Age of Jefferson to the Age of Ecology*. Chicago, IL: University of Chicago Press, 2009.

Baudrillard, Jean. *America*. New York: Verso, 1989.

———. *Simulacra and Simulation*. Translated by Sheila Farrier Glaser. Michigan: The University of Michigan Press, 1994.

Behe, Michael J. *The Edge of Evolution: The Search for the Limits of Darwinism*. New York: Free Press, 2007.

Bell, Daniel. *The Coming of Post-Industrial Society: A Venture in Social Forecasting*. New York: Basic Books, 1976.

Bergin, Mark. "Junk Science." *World Magazine* 21, no. 8 (February 25, 2006).

Bergreen, Laurence. *Over the Edge of the World: Magellan's Terrifying Circumnavigation of the Globe*. New York: Harper Perennial, 2003.

Bertlmann, R. A., and A. Zeilinger, eds. *Quantum [Un]speakables: From Bell to Quantum Information*. Berlin: Springer, 2002.

Bird, Kai, and Martin J. Sherwin. *American Prometheus: The Triumph and Tragedy of J. Robert Oppenheimer*. New York: Alfred A. Knopf, 2005.

Bizony, Piers. *One Giant Leap: Apollo 11 Remembered*. Minneapolis, MN: Zenith Press, 2009.

Blackwell, Richard J. *Behind the Scenes at Galileo's Trial*. Notre Dame, IN: University of Notre Dame Press, 2006.

Bluestone, Barry and Bennett Harrison. *The Deindustrialization of America: Plant Closings, Community Abandonment, and the Dismantling of Basic Industry*. New York: Basic Books, 1982.

Blumer, Herbert. *Symbolic Interactionism: Perspective and Method*. New York: Prentice Hall, 1969.

Blunt, Wilfred. *Linnaeus: The Compleat Naturalist*. London: Frances Lincoln Limited, 2004.

Bohr, Niels. "The Quantum Postulate and the Recent development of Atomic Theory." In *The Philosophical Writings of Niels Bohr: Volume 1–Atomic Theory and the Description of Nature*. Woodbridge, CT: Ox Bow Press, 1928.

———. "Can Quantum-Mechanical Description of Physical Reality Be Complete?" *Physical Review* 48 (1935): 696-702.

Born, Max. *The Born–Einstein Letters 1916-1955: Friendship, Politics and Physics in Uncertain Times*. New York: Macmillan, 2005.

Bose, Meena and Rosanna Perotti, eds. *From Cold War to New World Order: The Foreign Policy of George H. W. Bush*. Westport, CT: Greenwood Press, 2002.

Bourdieu, Pierre. *Distinction: A Social Critique of the Judgment of Taste*. Cambridge, MA: Harvard University Press, 1984.

Briggs, Asa, and Peter Burke. *Social History of the Media: From Gutenberg to the Internet*. Malden, MA: Polity Press, 2009.

Brighton, Henry, and Howard Selina. *Introducing Artificial Intelligence*. Cambridge, UK: Totem Books, 2003.

Brock, David C., ed. *Understanding Moore's Law: Four Decades of Innovation*. Philadelphia, PA: Chemical Heritage Press, 2006.

Brown, Colin. *Christianity and Western Thought: A History of Philosophers, Ideas and Movements. Volume 1: From the Ancient World to the Age of Enlightenment*. Downers Grove, IL: InterVarsity Press, 1990.

Buckland, William. *Reliquiae Diluvianae: or Observations on the Organic Remains Contained in Caves, Fissures and Diluvial Gravel, and on Other Geological Phenomena Attesting the Action of an Universal Deluge*. London: John Murray, 1824.

Burawoy, Michael. *Manufacturing Consent: Changes in the Labor Process Under Monopoly Capitalism*. Chicago: University of Chicago Press, 1979.

Cadbury, Deborah. *The Space Race: The Epic Battle Between America and the Soviet Union for the Dominion of Space*. New York: HarperCollins, 2006.

Carroll, Lewis. *Alice in Wonderland*. Boston: Barta Press, 1897.

Carroll, Michael. *The Seventh Landing: Going Back to the Moon: This Time to Stay*. New York: Springer, 2009.

Carter, Matt. *Minds and Computers: An Introduction to the Philosophy of Artificial Intelligence*. Edinburgh, Scotland: Edinburgh University Press, 2007.

Chaikin, Andrew. *Man on the Moon: The Voyages of the Apollo Astronauts*. New York: Penguin Books, 2007.

Clark, Andy. *Natural-Born Cyborgs: Minds, Technologies, and the Future of Human Intelligence*. Oxford: Oxford University Press, 2003.

Clegg, Brian. *The God Effect: Quantum Entanglement, Science's Strangest Phenomenon*. New York: MacMillan, 2006.

———. *Before the Big Bang: The Prehistory of Our Universe*. New York: St. Martin's Press, 2009.

Clough, Patricia Ticineto. *The End(s) of Ethnography: From Realism to Social Criticism*. Newbury Park: Sage, 1992.

———. *Feminist Thought: Desire, Power and Academic Discourse*. Cambridge, MA: Blackwell, 1994.

Collins, Francis S. *The Language of God: A Scientist Presents Evidence for Belief*. New York: Free Press, 2006.

Cooley, Charles Horton. *Human Nature and the Social Order*. New York: Charles Scribner's Sons, 1902.

Dahl, Robert A. "The Concept of Power." *Behavioural Science* 2 (1957): 201-205.

———. *Who Governs? Democracy and Power in an American City*. New Haven: Yale University Press, 1961.

Darwin, Charles. *Journal of researches into the natural history and geology of the countries visited during the voyage of H.M.S. Beagle round the world, under the Command of Capt. FitzRoy, R.N.* 2d edition, London: John Murray, 1845.

Darwin, Charles and Alfred Russel Wallace. "On the Tendency of Species to form Varieties; and on the Perpetuation of Varieties and Species by Natural Means of Selection." *Zoology* 3, Journal of the Proceedings of the Linnean Society of London (1858): 46–50.

———. *On the Origin of Species by Means of Natural Selection, or the Preservation of Favoured Races in the Struggle for Life*. 1st ed. London: John Murray, 1859.

———. *The Descent of Man and Selection in Relation to Sex*. Volume 1. New York: D. Appleton and Company, 1871.

———. *Life and Letters of Charles Darwin*. Vol. 1. New York: D. Appleton and Co, 1887.

———. *The Autobiography of Charles Darwin*, 1809-1882. Nora Barlow, ed. New York: W. W. Norton, 1958.

———. *The Autobiography of Charles Darwin*. United Kingdom: The Complete Works of Charles Darwin Online, 1990. www.darwin-online.org.uk/content/frameset?viewtype=side&itemID=F1497&pageseq=64/.

Darwin, Erasmus. *Zoonomia: The Laws of Organic Life*. London: J. Johnson, 1796.

Davis, Kingsley. "Extreme Social Isolation of a Child." *The American Journal of Sociology* 45, no. 4 (January, 1940): 554-565.

Dawkins, Richard. *The Greatest Show on Earth: The Evidence for Evolution*. New York: Free Press, 2009.

De Roy, Tui (ed.). *Galapagos: Preserving Darwin's Legacy*. New York: Longitude Books.

Denzin, Norman K., and Yvonna S. Lincoln. *Handbook of Qualitative Research*. Thousand Oaks: Sage Publications, 1994.

Denzin, Norman K. "The Standpoint Epistemologies and Social Theory." *Current Perspectives in Social Theory* 17 (1997): 39-76.

Derrida, Jacques. *Writing and Difference*. Translated by Alan Bass. Chicago: University of Chicago Press, 1978.

Descartes, Rene. *The Method, Meditations, and Philosophy of Descartes*. New York: Tudor, 1901.

Desmond, Adrian and James Moore. *Darwin's Sacred Cause: How a Hatred of Slavery Shaped Darwin's Views on Human Evolution*. Boston: Houghton Mifflin Harcourt, 2009.

Diamond, Jared M. *Guns, Germs, and Steel: The Fates of Human Societies*. New York: W.W. Norton & Co., 1998.

———. *Collapse: How Societies Choose to Fail or Succeed*. New York: Viking, 2005.

———. *The Third Chimpanzee: The Evolution and Future of the Human Animal*. New York: HarperCollins, 2006.

Dickson, Paul. *Sputnik: The Shock of the Century*. New York: Berkley Books, 2003.

Doak, Robin S. *Galileo: Astronomer and Physicist*. Minneapolis, MN: Compass Point Books, 2005.

Dobbs, David. *Reef Madness: Charles Darwin, Alexander Agassiz, and the Meaning of Coral*. New York: Pantheon, 2005.

Domhoff, G. William. *Who Rules America: Power and Politics, and Social Change*. New York: McGraw-Hill, 2006.

D'Onofrio, Mauro and Carlo Burigana, eds. *Questions of Modern Cosmology: Galileo's Legacy*. New York: Springer, 2009.

Dretzin, Rachel (Director and Producer). "Digital Nation: Life on the Virtual Frontier." *Frontline*. Boston: Public Broadcasting Service, WGBH, 2010.

Dreyfus, Hubert. *What Computers Still Can't Do: A Critique of Artificial Reason*. Cambridge, MA: MIT Press, 1992.

Dunsworth, Holly M. *Human Origins 101*. Westport CT: Greenwood Press., 2007.

Ehrlich, Paul R., and Ann H. Ehrlich. *The Population Explosion*. New York: Simon and Schuster, 1991.

Einstein, Albert, Boris Podolsky and Nathan Rosen. "Can Quantum-Mechanical Description of Physical Reality Be Considered Complete?" *Physical Review* 47 (1935): 777-780.

Einstein, Albert. *Out of My Later Years*. New York: Philosophical Library, 1950.

Eisenhower, Dwight D. *The Military-Industrial Complex: President Dwight D. Eisenhower's Farewell Address to the Nation*. With an Introduction by Jesse Smith (2006). Basementia Publications, 1961.

Erickson, Mark. *Into the Unknown Together: The DOD, NASA, and Early Spaceflight*. Maxwell Air Force Base, AL: Air University Press, 2005.

Feldhay, Rivka. *Galileo and the Church: Political Inquisition or Critical Dialogue?* Cambridge, UK: Cambridge University Press, 1995.

Finkbeiner, Ann. *A Grand and Bold Thing: An Extraordinary New Map of the Universe Ushering in a New Era of Discovery*. New York: Free Press, 2010.

Finocchiaro, Maurice A. *The Galileo Affair: A Documentary History*. Berkeley, CA: University of California Press, 1989.

FitzRoy, Robert. *Narrative of the Surveying Voyages of His Majesty's Ships Adventure and Beagle Between the Years 1826 and 1836, Describing Their Examination of the Southern Shores of South America, and the Beagle's Circumnavigation of the Globe. Proceedings of the Second Expedition, 1831-36, Under the Command of Captain Robert FitzRoy, R.N.* London: Henry Colburn, 1839.

Folsing, Albrecht. *Albert Einstein: A Biography*. London: Viking, 1997.

Foucault, Michel. *Discipline and Punish: The Birth of the Prison*. Translated by Alan Sheridan. New York: Pantheon Books, 1977.

———. *Power/Knowledge: Selected Interviews and Other Writings 1972-1977*. Edited by Colin Gordon. New York: Pantheon Books, 1980.

Fraser, Steve, ed. *The Bell Curve Wars: Race, Intelligence, and the Future of America*. New York: Basic Books, 1995.

Freeman, Milton M. R., and Urs P. Kreuter, eds. *Elephants and Whales: Resources for Whom?* Basel, Switzerland: Gordon and Breach Science Publishers, 1994.

French, Francis, and Colin Burgess. *In the Shadow of the Moon: A Challenging Journey to Tranquility, 1965-1969.* Lincoln: University of Nebraska Press, 2007.

Freud, Sigmund. *Civilization and Its Discontents.* New York: W. W. Norton, 1961.

Fried, Albert. *McCarthyism: The Great American Red Scare: A Documentary History.* Oxford: Oxford University Press, 1997.

Friedman, Thomas L. *Hot, Flat, and Crowded: Why We Need a Green Revolution—and How it Can Renew America.* New York: Farrar, Straus and Giroux, 2008.

Fripp, Jon, Michael Fripp and Deborah Fripp, eds. *Speaking of Science: Notable Quotes on Science, Engineering, and the Environment.* Eagle Rock, VA: LLH Technology Publishing, 2000.

Gaddis, John Lewis. *The Cold War: A New History.* New York: Penguin, 2005.

Galbraith, John Kenneth. *The Affluent Society.* Boston: Houghton Mifflin, 1958.

Gane, Mike. *Baudrillard: Critical and Fatal Theory.* New York: Routledge, 1991.

Garwood, Christine. *Flat Earth: The History of an Infamous Idea.* New York: Thomas Dunne Books, 2008.

Giddens, Anthony. *The Constitution of Society: Outline of the Theory of Structuration.* Berkeley: University of California Press, 1984.

———. "Riding the Juggernaut of Modernity." Pp. 39-42 in *Sociology: Introductory Readings.* 3rd Edition. Anthony Giddens, and Phillip W. Sutton, eds. Cambridge, UK: Polity Press, 2010.

Gingerich, Owen. *The Book Nobody Read: Chasing the Revolutions of Nicolaus Copernicus.* New York: Walker & Company, 2004.

Gould, Stephen Jay. *Wonderful Life: The Burgess Shale and the Nature of History.* New York: W. W. Norton, 1989.

———. *The Mismeasure of Man.* New York: W. W. Norton and Company, 1996.

———. *Full House: The Spread of Excellence from Plato to Darwin.* New York: Three Rivers Press, 1997.

Grant, K. Thalia, and Gregory B. Estes. *Darwin in Galapagos: Footsteps to a New World.* Princeton, NJ: Princeton University Press, 2009.

Greene, Brian. *The Elegant Universe: Superstrings, Hidden Dimensions, and the Quest for the Ultimate Theory.* New York: W. W. Norton and Company, 2003.

Habermas, Jürgen. "Toward a Theory of Communicative Competence." *Inquiry* 13 (1970): 360-365.

———. *Knowledge and Human Interests.* London: Heinemann Educational Books, 1972.

———. *Theory of Communicative Action.* 2 vols. London: Heinemann. Vol. 1, *Reason and the Rationalization of Society*, also published by Polity Press, Cambridge, England, 1981.

Hankins, Thomas L. *Science and the Enlightenment.* Cambridge, UK: University of Cambridge Press, 1985.

Harland, David M. *The First Men on the Moon: The Story of Apollo 11.* New York: Springer, 2007.

Hawking, Stephen W. *A Brief History of Time: From the Big Bang to Black Holes.* New York: Bantam Books, 1988.

———. *The Theory of Everything: The Origin and Fate of the Universe.* Beverly Hills, CA: New Millennium Press, 2002.

Heisenberg, Werner. *The Physical Principles of the Quantum Theory.* Translated by Carl Eckart and F. C. Hoyt. Chicago: University of Chicago Press, 1930.

Henneberg, Maciej and John Schofield. *The Hobbit Trap: Money, Fame, Science and the Discovery of a 'New Species'.* Kent Town: Wakefield Press, 2008.

Herrnstein, Richard J., and Charles Murray. *The Bell Curve: Intelligence and Class Structure in American Life*. New York: The Free Press, 1994.

Hoffman, Paul. *King's Gambit: A Son, A Father, and the World's Most Dangerous Game*. New York: Hyperion, 2007.

Hofstadter, Dan. *The Earth Moves: Galileo and the Roman Inquisition*. New York: W. W. Norton, 2009.

Hofstadter, Douglas, R. *Gödel, Escher, Bach: An Eternal Golden Braid*. New York: Basic Books, 1979.

Hogan, Michael J. *The Marshall Plan: America, Britain, and the Reconstruction of Western Europe, 1947-1952*. Cambridge, UK: Cambridge University Press, 1987.

Hornblum, Allen M. *The Invisible Harry Gold: The Man Who Gave the Soviets the Atom Bomb*. New Haven, CT: Yale University Press, 2010.

Horner, Jack, and James Gorman. *How to Build a Dinosaur: Extinction Doesn't Have to be Forever*. New York: Dutton, 2009.

Houghton, John. *Global Warming: The Complete Briefing*. Third Edition. Cambridge, UK: Cambridge University Press, 2004.

Hsu, Tiffany. "NYU Art Professor to Take Photos with Camera Implanted in the Back of His Head." *LA Times*, December 6, 2010. www.latimesblogs.latimes.com/technology/2010/12/nyu-art-professor-to-take-photos-with-camera-implanted-in-the-back-of-his-head.html

Hunt, Earl. *Human Intelligence*. Cambridge, UK: Cambridge University Press, 2010.

Isaacson, Walter. *Einstein: His Life and Universe*. New York: Simon & Schuster, 2007.

Jacoby, Russell and Naomi Glauberman, eds. *The Bell Curve Debate: History, Documents, Opinions*. New York: Times Books, 1995.

Johnston, Sean F. *History of Science: A Beginner's Guide*. Oxford: One World, 2009.

Kaku, Michio. *Einstein's Cosmos: How Albert Einstein's Vision Transformed Our Understanding of Space and Time*. New York: W. W. Norton, 2004.

———. *Physics of the Impossible: A Scientific Exploration Into the World of Phasers, Force Fields, Teleportation, and Time Travel*. New York: Doubleday Publishing, 2008.

———. *The Physics of the Future: How Science Will Shape Human Destiny and Our Daily Lives by the Year 2100*. New York: Knopf Doubleday, 2011.

Keynes, Richard Darwin. *Darwin, His Daughter, and Human Evolution*. London: Riverhead Books, 2002.

———. *Fossils, Finches, and Fuegians: Darwin's Adventures and Discoveries on the Beagle*. Oxford: Oxford University Press, 2003.

Kincheloe, Joe L. and Peter L. McLaren, "Rethinking Critical Theory and Qualitative Research." Pp. 138-157 in *Handbook of Qualitative Research*. Norman K. Denzin and Yvonna S. Lincoln, eds. Thousand Oaks, CA: Sage, 1994.

Kincheloe, Joe L., Shirley R. Steinberg, and Aaron D. Gresson. *Measured Lies: The Bell Curve Examined*. New York: St. Martin's Press, 1997.

King, Henry C. *The History of the Telescope*. New York: Charles Griffin and Company, Ltd., 1955.

Koeberl, Christian and Kenneth G. Macleod, eds. *Catastrophic Events and Mass Extinctions: Impacts and Beyond*. Boulder, CO: The Geological Society of America, Inc., 2002.

Kuhn, Thomas S. *The Structure of Scientific Revolutions*. Chicago: University of Chicago Press, 1962.

———. *The Structure of Scientific Revolutions*. Third Edition. Chicago: University of Chicago Press, 1996.

Kumar, Manjit. *Quantum: Einstein, Bohr, and the Great Debate About the Nature of Reality*. New York: W. W. Norton, 2008.

Kurzweil, Ray. *The Age of Intelligent Machines*. Cambridge, MA: MIT Press, 1992.

———. *The Age of Spiritual Machines: When Computers Exceed Human Intelligence*. New York: Penguin, 1999.

———. *The Singularity is Near: When Humans Transcend Biology*. New York: Viking, 2005.

Ladyman, James. *Understanding Philosophy of Science*. New York: Routledge, 2002.

Lakatos, Imre and Alan Musgrave, eds. *Criticism and the Growth of Knowledge: Volume 4: Proceedings of the International Colloquium in the Philosophy of Science*. London: Cambridge University Press, 1970.

Landsberger, Henry A. *Hawthorne Revisited: Management and the Worker: Its Critics, and Developments in Human Relations in Industry*. Ithaca, NY: Cornell University Press, 1958.

Larsen, Kristine M. *Stephen Hawking: A Biography*. Westport, CT: Greenwood Press, 2005.

Lather, Patti. *Getting Smart: Feminist Research and Pedagogy With/in the Postmodern*. New York: Routledge, 1991.

———. "The Validity of Angels: Interpretive and Textual Strategies in Researching the Lives of Women With HIV/AIDS." *Qualitative Inquiry* 1, no. 1 (1995): 41-68.

Lemert, Charles. "The End of Ideology, Really." *Sociological Theory* 9, no. 2 (Fall, 1991): 164-172.

———, ed. *Social Theory: The Multicultural and Classic Readings*. 3rd Edition. Boulder, CO: Westview Press, 2004.

Lenski, Gerhard E. *Power and Privilege: A Theory of Stratification*. New York: McGraw-Hill, 1966.

Lenski, Gerhard and Patrick Nolan. *Human Societies: An Introduction to Macrosociology*. 10th Edition. New York: Paradigm Publishers, 2006.

Lindroth, Sten. "The Two Faces of Linnaeus," in *Linnaeus: The Man and His Work*, Tore Frangsmyr, ed. Berkeley: University of California Press, 1983.

Lipset, Seymour Martin and Reinhart Bendix. *Social Mobility in Industrial Society*. New Brunswick, NJ: Transaction Publishers, 1992.

Louv, Richard. *The Last Child in the Woods: Saving Our Children from Nature Deficit Disorder*. Chapel Hill, NC: Algonquin Books of Chapel Hill, 2008.

Lovejoy, Arthur Oncken. *The Great Chain of Being: A Study of the History of an Idea*. Boston: Harvard University Press, 1936.

Lukes, Steven. *Power: A Radical View*. Second Edition. London: Palgrave Macmillan, 2005.

Lyell, Charles. *Principles of Geology: Being an Attempt to Explain the Former Changes of the Earth's Surface, By References to Causes Now in Operation*. Volume One. London: William Clowes, 1832.

Lyng, Stephen and David D. Franks. *Sociology and the Real World*. Lanham, MD: Rowman & Littlefield, 2002.

Lyotard, Jean-Francois. *The Postmodern Condition: A Report on Knowledge*. Translated by Geoff Bennington and Brian Massumi. Minneapolis: University of Minnesota Press, 1984.

Lynch, Gary and Richard Granger. *Big Brain: The Origins and Future of Human Intelligence*. New York: Palgrave Macmillan, 2008.

Malthus, Thomas. *An Essay On The Principle Of Population* (1798 1st edition, plus excerpts 1803 2nd edition). New York: W. W. Norton, 2003.

Marx, Karl, and John C. Raines. *Marx on Religion*. Philadelphia, PA: Temple University Press, 2004.

Mawer, Simon. *Gregor Mendel: Planting the Seeds of Genetics*. Chicago, IL: Field Museum of Natural History, 2006.

McCorduck, Pamela. *Machines Who Think*. Natick, MA: A.K. Peters, Ltd., 2004.

McGettigan, Timothy. "Uncorrected Insight: Metaphor and Transcendence 'After Truth' in Qualitative Inquiry." *Qualitative Inquiry* 3, no. 3 (1997): 366-383.

———. "Redefining Reality: Epiphany as a Standard of Postmodern Truth." *Electronic Journal of Sociology* 3, no. 4 (1998). www.sociology.org/vol003.004/mcgettigan.article.1998.html

———. *Utopia on Wheels: Blundering Down the Road to Reality*. Lanham, MD: University Press of America, 1999.

———. "Flawed by Design: The Virtues and Limitations of Postmodern Theory." *Theory & Science* 1, no. 1, 2000. http://theoryandscience.icaap.org/content/vol001.001/-05mcgettigan.html.

———. "Field Research for Boneheads: From Naïveté to Insight on the Green Tortoise." *Sociological Research Online* 6, no. 2 (2001). www.socresonline.org.uk/6/2/mcget tigan.html.

———. "Redefining Reality: A Resolution to the Paradox of Emancipation and the Agency-Structure Dichotomy." *Theory & Science* 3, no. 2 (2002). http://theoryand science.icaap.org/volume3issue2.php.

———. "Redefining Reality: A Resolution to Theoretical Paradox in the Field of Sociology." *Electronic Journal of Sociology*, Tier One, 2006. www.sociology.org/content/2006/tier1/mcGettigan_RedefiningReality.pdf.

———. "Anomaly Overload: An Evolutionary Theory of truth." *Theory & Science* 10, no. 1 (2008). theoryandscience.icaap.org/content/vol10.1/ McGettigan1.html.

———. "May the Fittest Survive: The National Academy of Sciences vs. Creationism." A Review of Science, Evolution, and Creationism. Washington, DC: The National Academies Press. *The Socjournal*, September 14, 2010. www.sociology.org/book-reviews/fittest-survive-national-academy-sciences-creationism/.

McLaren, Peter. *Che Guevara, Paulo Freire, and the Pedagogy of the Revolution*. Lanham, MD: Rowman & Littlefield, 2000.

McLean, Bethany, and Peter Elkind. *The Smartest Guys in the Room: The Amazing Rise and Scandalous Fall of Enron*. New York: Portfolio, 2004.

McNeese, Tim. *Christopher Columbus and the Discovery of the Americas*. New York: Chelsea House Publishers, 2006.

Mead, George Herbert. *Mind, Self, and Society*. Edited by C. W. Morris. Chicago: University of Chicago Press, 1934.

Meier, Deborah, Alfie Kohn, Linda Darling-Hammond, Theodore R. Sizer, and George Wood. *Many Children Left Behind: How the No Child Left Behind Act is Damaging Our Children and Our Schools*. Boston: Beacon Press, 2004.

Mills, C. Wright. *The Power Elite*. Oxford: Oxford University Press, 1956.

———. *The Sociological Imagination*. New York: Oxford University Press, 1959.

Minow, Newton N. "Television and the Public Interest." Address to the National Association of Broadcasters, Washington, DC, May 9, 1961.

Minsky, Marvin. *The Society of Mind*. New York: Simon and Schuster, 1985.

———. *The Emotion Machine: Commonsense Thinking, Artificial Intelligence, and the future of the Human Mind*. New York: Simon and Schuster, 2006.

Montandon, Mac. *Jetpack Dreams: One Man's Up and Down (But Mostly Down) Search for the Greatest Invention that Never Was*. New York: Da Capo Press, 2008.

Morison, Samuel Eliot. *Admiral of the Ocean Sea: A Life of Christopher Columbus*. Camden, NJ: Little, Brown and Company, 1942.

Murray, Charles. *Losing Ground: American Social Policy, 1950-1980*. New York: Basic Books, 1984.

———. *Income Inequality and IQ*. Washington, DC: AEI Press, 1998.

Nahin, Paul J. *Time Machines: Time Travel in Physics, Metaphysics, and Science Fiction*. New York: Springer-Verlag, 1999.

National Academy of Science. *Science, Evolution, and Creationism*. Washington, DC: National Academies Press, 2008.

Nelson, Richard William. *Darwin, Then and Now: The Most Amazing Story in the History of Science*. Bloomington, IN: iUniverse, 2009.

Newborn, Monty. *Deep Blue: An Artificial Intelligence Milestone*. New York: Springer Verlag, 2003.

Nilsson, Nils J. *The Quest for Artificial Intelligence: A History of Ideas and Achievements*. Cambridge, UK: Cambridge University Press, 2010.

Nof, Shimon Y. *Handbook of Industrial Robotics*. Volume 1. New York: Wiley and Sons, 1999.

Norris, Robert S. and William M. Arkin. "Russian/Soviet Weapons Secretes Revealed." *Bulletin of the Atomic Scientists* 49, no. 3 (1993).

Ohanian, Hans C. *Einstein's Mistakes: The Human Failings of Genius*. New York: W.W. Norton & Company, 2008.

Okin, J. R. *The Technology Revolution: The Not-For-Dummies Guide to the Impact, Perils, and Promise of the Internet*. Winter Harbor, ME: Ironbound Press, 2005.

Orloff, Richard W., and David M Harland. *Apollo: The Definitive Sourcebook*. New York: Springer, 2006.

Outram, Dorinda. *The Enlightenment*. Second Edition. Cambridge, UK: Cambridge University Press, 2005.

Overbye, Dennis. "Knowing the Universe in Detail (Except for That Pesky 96 Percent of It)." *New York Times*, October 24, 2006. www.nytimes.com/2006/10/24/science/space/24essa.html.

Owen, David. *Final Frontier: Voyages into Outer Space*. Buffalo, NY: Firefly Books, 2004.

Pais, Abraham. *J. Robert Oppenheimer: A Life*. Oxford: Oxford University Press, 2006.

Parsons, Talcott. *Structure and Process in Modern Societies*. Glencoe, IL: Free Press, 1960.

Piaget, Jean. *The Psychology of Intelligence*. London: Routledge and Kegan Paul, 1951.

Pollan, Michael. *The Omnivore's Dilemma: A Natural History of Four Meals*. New York: Penguin Press, 2006.

Popper, Karl. *Unended Quest: An Intellectual Biography*. New York: Routledge, 2002.

———. *All Life is Problem Solving*. Translated by Patrick Camiller. New York: Routledge, 1999.

———. *In Search of a Better World: Lectures and Essays from Thirty Years*. Translated by Laura J. Bennett. New York: Routledge, 1992.

———. *The Open Universe: An Argument for Indeterminism*. New York: Routledge, 1988.

———. *Realism and the Aim of Science*. Totowa, NJ: Rowman & Littlefield, 1983.

———. *The Open Universe: An Argument for Indeterminism*. Totowa, NJ: Rowman & Littlefield, 1982a.

———. *Quantum Theory and the Schism in Physics*. Totowa, NJ: Rowman & Littlefield, 1982b.

———. *Objective Knowledge: An Evolutionary Approach*. London: Oxford University Press, 1972.

———. *Conjectures and Refutations: The Growth of Scientific Knowledge*. New York: Basic Books, 1965.

———. *The Logic of Scientific Discovery*. New York: Basic Books, 1959.

———. *The Poverty of Historicism*. New York: Harper Torchbooks, 1957.

———. *The Open Society and its Enemies*. Princeton, NJ: Princeton University Press, 1950.

Quammen, David. *The Reluctant Mr. Darwin: An Intimate Portrait of Charles Darwin and the Making of his Theory of Evolution*. New York: Atlas Books/Norton, 2006.

Ray, John. *Wisdom of God Manifested in the Works of the Creation*. London: Printed by R. Harbin, for William Innys, at the Prince's-Arms in St Paul's Church Yard, 1691.

Reader, John. "Whatever Happened to Zinjantropus?" *New Scientist*, 26 (March, 1981): 802-805.

Reisner, Marc. *Cadillac Desert: The American West and its Disappearing Water*. New York: Penguin Books, 1993.

Repcheck, Jack. *Copernicus' Secret: How the Scientific Revolution Began*. New York: Simon and Schuster, 2007.

Rhodes, Richard. *The Making of the Atomic Bomb*. New York: Simon & Schuster, 1986.

———. *Dark Sun: The Making of the Hydrogen Bomb*. New York: Simon & Schuster, 1995.

———. *Arsenals of Folly: The Making of the Nuclear Arms Race*. New York: Alfred A. Knopf, 2007.

Richardson, Laurel. "Postmodern Social Theory: Representational Practices." *Sociological Theory* 9, no. 2 (Fall, 1991): 173-179.

Ritchie, Nick. *US Nuclear Weapons Policy After the Cold War: Russians, 'Rogues,' and Domestic Division*. New York: Routledge, 2009.

Ritzer, George. *Sociological Theory*. Fifth Edition. New York: McGraw Hill, 2000.

———. *The McDonaldization of Society 5*. Thousand Oaks, CA: Pine Forge Press, 2008.

Rostow, Walt W. *The Stages of Economic Growth: A Non-Communist Manifesto*. Cambridge: Cambridge University Press, 1960.

Rufus, Annali and Kristan Lawson. *The Scavengers' Manifesto*. New York: Penguin, 2009.

Schefter, James. *The Race: The Uncensored Story of How America Beat Russia to the Moon*. New York: Doubleday, 1999.

Schlosser, Eric. *Fast Food Nation: The Dark Side of the All-American Meal*. New York: Houghton Mifflin, 2001.

Schumpeter, Joseph. *Capitalism, Socialism and Democracy*. New York: Harper, 1975.

Searle, John. "Minds, Brains and Programs." *Behavioral and Brain Sciences* 3, no. 3 (1980): 417–457.

Seelhorst, Mary. "Jules Verne." *Popular Mechanics* (July, 2003): 36-37.

Seidman, Steven. "The End of Sociological Theory: The Postmodern Hope." *Sociological Theory* 9, no. 2 (Fall, 1991): 131-146.

Shulman, Seth. *Unlocking the Sky: Glenn Hammond Curtiss and the Race to Invent the Airplane*. New York: HarperCollins, 2003.

Smith, Dorothy E. *The Conceptual Practices of Power: A Feminist Sociology of Knowledge*. London: Routledge, 1990.

Smout, Kary D. *The Creation/Evolution Controversy: A Battle for Cultural Power*. New York: Praeger, 1998.

Sobel, Dava. *Longitude: The True Story of a Lone Genius Who Solved the Greatest Scientific Problem of His Time.* New York: Walker Publishing Company, 1995.

———. *Galileo's Daughter: A Historical Memoir of Science, Faith, and Love.* New York: Penguin Books, 2000.

Sokal, Alan. "A Physicist Experiments with Cultural Studies," *Lingua Franca* (May/June 1996).

———. *Beyond the Hoax: Science, Philosophy and Culture.* Oxford: Oxford University Press, 2010.

Spurlock, Morgan. *Super-Size Me.* Kathbur Pictures, 2004.

Steele, Phillip. *Galileo: The Genius Who Faced the Inquisition.* Washington, DC: National Geographic Society, 2008.

Sunderland, Luther. *Darwin's Enigma: Ebbing the Tide of Naturalism.* Green Forest, AZ: Master Books, 1988.

Swade, Doron. *The Difference Engine: Charles Babbage and the Quest to Build the first Computer.* New York: Viking Press, 2001.

Swedin, Eric G., and David L. Ferro. *Computers: The Life Story of a Technology.* Baltimore, MD: Johns Hopkins Press, 2007.

Taylor, Frederick Winslow. *The Principles of Scientific Management.* New York: Harper and Brothers Publishers, 1911.

Teller, Edward, and Judith Schoolery. *Memoirs: A Twentieth Century Journey in Science and Politics.* New York: Perseus, 2001.

Thimmesh, Catherine. *Team Moon: How 400,000 People Landed Apollo 11 on the Moon.* New York: Houghton Mifflin, 2006.

Thompson, Clive. "Smarter Than You Think: What is IBM's Watson?" *New York Times,* June 16, 2010. www.nytimes.com/2010/06/20/magazine/ 20Computer-t.html?_r=1

Thompson, Herb. "Cybersystemic Learning." *Radical Pedagogy* 3, no. 2 (2001). http://radical-pedagogy.icaap.org/content/issue3_2/thompson.html.

Tierney, William G. *Academic Outlaws: Queer Theory and Cultural Studies in the Academy.* Thousand Oaks, CA: Sage, 1997.

Tise, Larry E. *Conquering the Sky: The Secret Flights of the Wright Brothers at Kitty Hawk.* New York: Palgrave Macmillan, 2009.

Tobin, James. *To Conquer The Air: The Wright Brothers and the Great Race for Flight.* New York: Simon & Schuster, 2004.

Toffler, Alvin. *Future Shock.* New York: Random House, 1970.

Turing, A.M. "Computing Machinery and Intelligence." *Mind,* 59 (1950): 433-460.

Turse, Nick. *The Complex: How the Military Invades Our Everyday Lives.* New York: Metropolitan Books, 2008.

Van Dorn, W. G. *Ivy Mike: The First Hydrogen Bomb.* Bloomington, IN: Xlibris, 2008.

Venneti, Nancy R. *Labor, Job Growth, and the Workplace of the Future.* Hauppauge, NY: Nova Science Publishers, 2003.

Watson, James D. *The Double Helix: A Personal Account of the Discovery of the Structure of DNA.* New York: Scribner, 1968.

Webster, Frank. *Theories of the Information Society.* New York: Routledge, 1995.

Weizenbaum, Joseph. *Computer Power and Human Reason: From Judgment to Calculation.* New York: W. H. Freeman and Company, 1976.

Willis, Paul. *Learning to Labor: How Working Class Kids Get Working Class Jobs.* New York: Columbia University Press, 1977.

Wolfson, Richard. *Simply Einstein: Relativity Demystified.* New York: W. W. Norton, 2003.

Wittgenstein, Ludwig. *Tractatus Logico-Philosophicus.* London: Routledge & Paul, 1961.

Wright, Orville. *How We Invented the Airplane: An Illustrated History.* Mineola, NY: Dover Publications, 1998.

Wright, Will. *Wild Knowledge: Science, Language, and Social Life in a Fragile Environment.* Minneapolis: University of Minnesota Press, 1992.

Index

About the Author

Timothy McGettigan is professor of sociology at Colorado State University–Pueblo. His interests include science, technology, and the future. His next book will examine the goals and limitations of scientific cosmology.